THE SCIENCE OF SUPERHEROES

THE SCIENCE OF SUPERHEROES

THE SECRETS BEHIND SPEED, STRENGTH, FLIGHT, EVOLUTION, AND MORE

MARK BRAKE

Racehorse Publishing

Racehorse Publishing books may be purchased in bulk at special discounts for sales promotion, corporate gifts, fund-raising, or educational purposes. Special editions can also be created to specifications. For details, contact the Special Sales Department, Skyhorse Publishing, 307 West 36th Street, 11th Floor, New York, NY 10018 or info@skyhorsepublishing.com.

Racehorse Publishing™ is a pending trademark of Skyhorse Publishing, Inc.®, a Delaware corporation.

Visit our website at www.skyhorsepublishing.com.

10 9 8 7 6 5 4 3 2 1

Library of Congress Cataloging-in-Publication Data is available on file.

Cover design by Michael Short
Cover photo by iStockphoto

Print ISBN: 978-1-63158-211-0
Ebook ISBN: 978-1-63158-213-4

Printed in the United States of America

For James

"Come with me now, my son. As we break through the bounds of your earthly confinement, traveling through time and space. Your powers will far exceed those of mortal men."

—*Superman Returns*

TABLE OF CONTENTS

PART I: SPACE

PART II: TIME

PART III: MACHINE

PART IV: MONSTER

INTRODUCTION

Superhero science fiction is turbo-charged and rammed full of ideas. Whether it's comic books or movies, superhero stories present us with a bewildering diversity of contrasting themes: extraordinary powers and masked vigilantes, crime and supervillainy, evolution and mutation, cosmic mythology, and alternate timelines. And that's just a sample! But on a more thoughtful level, we could identify four main themes: space, time, machine, and monster. Each of these themes is a way of exploring the relationship between the superhero fiction and the science and, beneath that, a deeper underlying theme of how we humans relate to the nonhuman nature of the science and technology of our ever-evolving universe. Taking a closer look at these themes will explain how this book is structured.

SPACE

The space theme of superhero fiction is usually some facet of the natural world, such as the nine realms of Norse mythology in Thor, or the genre's myriad alien creatures, which can be considered an animated version of nature. The seemingly incessant alien invasions of Earth fall under this theme, as do aspects of the physics of flight and motion.

TIME

This theme is about some kind of flux in the human condition, which is brought about by a process revealed in time, such as evolution. Tales on the topic of time often focus on the dialectic of history, so alternate timelines are to be found here, as are evolutionary fables such as immortality, the evolution of Aquaman, and the posthuman *homo superior*, Captain America.

MACHINE

All science fiction stories are jam-packed with machines, and superhero fiction is no different. Machine tales deal with the man-machine motif, including enhanced humans such as Cyborg and the creation of superweapons such as Iron Man's suit, Thor's hammer, and the shield of Captain America.

MONSTER

These superhero stories feature the nonhuman in the form of mutant or monster situated within humanity itself. In these tales there is often an agency of change such as a nuclear catastrophe, which leads to the change of human into nonhuman. It is within this theme that the remaking of man through genetic design is often encountered. Of course, monsters can be downbeat too, as the countless cases of supervillains testify.

Thinking about science fiction as the human versus the nonhuman is satisfyingly elegant and transparent. Mark Rose, professor in the English department at the University of California, Santa Barbara, deserves credit here. His theme of space, time, machine, and monster splendidly serves the purpose of charting superhero fiction's ongoing dialogue with science, so this book is structured around these four themes. It will allow us a closer look at how superhero fiction works in its special relationship with science.

PART I
SPACE

WHAT IF SUPERMAN PITCHED FOR THE METROPOLIS METEORS?

Exterior. Kent farm—daytime.

Clark is three now. He faces Jonathan, ten feet away. Like any three-year-old, Clark tosses a baseball to his dad. A few other baseballs scattered about.

JONATHAN: "Good. Excellent—nice arm."

Jonathan rolls the ball back to the kid. Clark picks it up, throws it again.

JONATHAN (cont'd): "Yes. Great control— you see that?"

CLARK: . . . yeah . . .

JONATHAN (tosses it back): "Okay, give it a little more juice. A little of that 'Kent magic.'"

And Clark hurls the ball— OUT OF THE STATE. Jonathan just watches it go.

JONATHAN (cont'd): . . . oh God . . .

—J. J. Abrams, *Superman* screenplay FIRST DRAFT (2002)

"There does seem to be a sense in which physics has gone beyond what human intuition can understand. We shouldn't be too surprised about that because we're evolved to understand things that move at a medium pace at a medium scale."

—Richard Dawkins, *The Very Best of Richard Dawkins: Quotes from a Devout Atheist* (2015)

"It followed from the special theory of relativity that mass and energy are both but different manifestations of the same thing—a somewhat unfamiliar conception for the average mind. Furthermore, the equation $E=mc^2$ in which energy is put equal to mass multiplied by the square of the velocity of light, showed that very small amounts of mass may be converted into a very large amount of energy and vice versa. The mass and energy were in fact equivalent, according to the formula mentioned before.

—Albert Einstein, quoted in *Atomic Physics*, by the J. Arthur Rank Organization, Ltd *(1948)*

"I didn't mean to hit the umpire with the dirt, but I did mean to hit that bastard in the stands."

—Babe Ruth, *New York Times (1922)*

"I'm not particularly fond of Gotham. It's like someone built a nightmare out of metal and stone."

—Superman, *Superman/Batman Vol 1, #53 (2008)*

BATMAN VERSUS SUPERMAN (MY TAKE ON A DELETED SCENE)

A full house at the Gotham stadium roars as the players take the pitch. Against a sweltering Gotham skyline, the green gauze of the baseball diamond is lit up for a steaming nighttime encounter: their very own Gotham Knights against fierce rivals, the Metropolis Meteors. Each side is playing its own superhero ringer. The Meteors are fielding Superman, and the Knights have high hopes that their home batter's box will feature the Batman himself. His slugging potential must surely be awesome. After all, his name *is* "bat-man," right?

Has there ever been a more dramatic debut? Not in baseball, anyhow. The competition is keen and the air electrified with tension. Men are born for games. Little else, it sometimes seems. Each one knows that play is nobler than work. He knows too that the value of a game is not innate to the game itself. Rather, it is in the value of that which is put at risk. Games of chance need a wager to warrant any meaning at all. Games of sport not only involve the skill and strength of the opponents, but also their humiliation in defeat. By trial of chance, or trial of worth, all games aspire to the state of war, for in play that which is wagered gulps up game, player, and all.

At the allotted time the lights seem to brighten further in the heat. The sheer din dies down. Superman steps forward to take his place on the pitcher's mound. Upon seeing the Man of Steel, a strange wall of noise sweeps across the crowd like a Mexican wave. Some start to catcall. Some rise from their seats and jeer, the heat and tension of the night getting the better of their sense of fair play. Expectation is high, yet what happens next takes everyone by surprise.

Superman pitches at such speed, normal everyday physics breaks down. From the back row of Gotham stadium it's simply impossible to tell, but Superman pitches the baseball at a pace approaching the speed of light, around 90 percent. The ball is moving so speedily that all else seems stationary. Even the particles and dust in the balmy Gothamite air stand still. Like the crowd, the air particles become passive observers of the ball's meteoric flight. Sure, the particles of air still vibrate at a few hundred miles an hour, but that super-pitched ball is passing through at 186,000 miles a second. As far as that ball is concerned, those particles are simply just hanging there, frozen, like microscopic pieces of stadium popcorn.

All that's aerodynamic melts into air. The usual rules simply don't apply. Normally, as you sometimes see illustrated on car commercials, the air would flow easily around an object speeding through it, but now it's as though Superman's pitch has frightened the particles of air into inaction. They don't have time to be bundled out of the way by the speeding baseball. Rather, the ball ploughs into them so hard that the atoms in the air molecules are either forced to fuse with the surface atoms of the approaching ball or simply pass straight through it.

Each minuscule collision that manages to occur issues a tiny explosion of scattered particles and gamma rays. Faster than the mind can see, but a perfect picture for a graphic novel, the expanding globule of gamma rays and debris radiates out from the vicinity of the pitcher's mound. The rays and debris start to rip apart the particles in the air, transforming Gotham stadium into a mushrooming bubble of radiant plasma. The leading wall of the bubble speedily approaches Batman the batsman at the speed of light, only infinitesimally ahead of the actual ball.

A matter of mere nanoseconds later, seventy to be exact, Superman's pitched ball arrives at Batman. Solidly planted at the home plate, Bats is not even sure he saw Superman let go of the baseball, as the light beam carrying that message would arrive at the same time as the ball itself. The ball arrives at the plate, eaten away by collisions along its short and swift journey. The ball is, by now, a rather appropriate bullet-shaped cloud of mushrooming plasma in the steamy Gotham night. It jets through the air as a shell of X-ray radiation hits the batsman first, instantly followed by the cloud of debris.

But Batman seems to have been the only one in the stadium to expect such a pitch. Hardly by wit, and far more likely by fate, Batman makes contact with the "ball." The bullet-shaped cloud of the ball is still moving at near light speed. First it hits the bat, but immediately Batman, plate, and catcher are all engulfed by the surging cloud and carried through the backstop as they all begin to degenerate into smithereens. The expanding sphere of radiation and white-hot plasma opens out in all directions, save the ground itself, as the cloud gobbles up the remaining players on the pitch, Gotham Stadium, and the city backstreets, as the stadium clock has failed to even tick a full second since the ball rocketed out of Superman's grip.

The scene cuts to some casual observers far outside the stadium, looking down on the city from a tall tower on the outskirts of Gotham. First, they see the blinding light, burning brighter than the midday sun. Then, seconds later, a mushrooming fireball morphs into a pyrocumulus cloud of devastation. They hear a terrifying roar as the baseball's blast wave approaches, ripping up road-signs and turning houses to tinder. They look on helplessly as the stadium neighborhood is leveled, and a firestorm starts to consume much of the city below. Out of sight, the baseball diamond is reduced to the kind of crater easily seen on the surface of the moon, which looks silently down on Gotham, a mere light second away.

Meanwhile, a dark speck is spied against the glow of the moon. Is it a bird? Is it a plane?

LIFT AND SEPARATE: THE CHALLENGE OF MAKING A "METEORITE" OUT OF SOKOVIA

"Sokovia, officially the Republic of Sokovia, is a small [fictional] country located in eastern Europe. Its capital city is Novi Grad. The tiny country was thrown further into disarray when Ultron implemented a plan to create a global extinction event by raising an area of land with a major city atop into the air and ramming it back into the ground. The plan was thwarted, primarily by the Avengers and their allies, along with limited assistance from the Sokovian police, but the portion of land lifted in the air ended up destroyed."

—Sokovia entry in the *Marvel Database*

ULTRON: [to the Vision] "You shut me out! You think I care? You take away my world, I take away yours."
He activates the Vibranium core and the Earth around Sokovia starts to shake and break.
TONY STARK: "Friday?"
As Sokovia is being destroyed.
FRIDAY: "Sokovia's going for a ride."
As the earth is shaking, falling in around them.
ULTRON: "Do you see? The beauty of it, the inevitability. You rise, only to fall. You, Avengers, you are my meteor, my swift and terrible sword, and the Earth will crack with the weight of your failure. Purge me from your computers, turn my own flesh against me. It means nothing. When the dust settles, the only thing living in this world will be metal."

—Joss Whedon, *Avengers: Age of Ultron* screenplay (2015)

TONY STARK: [Searching for secret door] "Please be a secret door, please be a secret door, please be a secret door . . . [finds a secret door in the Sokovian castle] Yay!"

—Joss Whedon, *Avengers: Age of Ultron* screenplay (2015)

"Propulsion is the force by which something, such as a ship, a car, or a space rocket, is moved forward. A rocket is propelled using slow-burning gas, which escapes through a nozzle (the narrow, back end of a rocket) to create a large amount of thrust and push the rocket upward. Crossing the 100km (62 miles) into space is easy enough. Space travel may be rocket science, but it isn't too tricky. You can get into space with a rocket the size of a telegraph pole, just by going quite fast and steering upward. The trick is staying up there. To stop yourself, and your spacecraft, from falling back to the Earth, you have to travel along your orbital path really, really quickly. To stay locked on your orbit, you need to set the craft's speed controls at about 5 miles per second."

—Mark Brake, *How To Be a Space Explorer: Your Out-of-this-World Adventure* (2014)

In the early years of this uncertain twenty-first century, jaw-dropping CGI cinematic science fiction seems to be almost everyone's thing. At the time of writing, the genre dominates the top fifteen highest-grossing movies of all time, with *Avatar, Star Wars: The Force Awakens, Jurassic World, The Avengers, Avengers: Age of Ultron, Iron Man 3,* and *Captain America: Civil War* all featuring in that list.

From *Avatar* to *Age of Ultron*, we all expect computer-generated imagery to loom large in big-budget blockbusters—it's stage center, right there up on the big silver screen. Some CGI is obvious. The appearance of the late Peter Cushing, digitally reanimated for *Rogue One: A Star Wars Story*, with the magic of cinema putting words into the mouth of a deceased actor. Then there's the simulation of iconic film-star footage of Elvis Presley and Marilyn Monroe in the epic futuristic fight-sequence of *Blade Runner 2049*.

Some CGI is about what you *can't* see, rather than what you can. Editors on *Justice League* had to remove a mustache grown by Superman actor, Henry Cavill, for his role in *Mission: Impossible 6*. Armie Hammer's genitals had to be digitally edited out of the movie romance *Call Me By Your Name*, as the actor's shorts had failed to fully contain his talents. And legend has it that one poor animator on the pig movie *Babe* spent months painstakingly removing every frame of the title character's anus. But it'll take some time before CGI produces another scene as jaw-dropping as the incredibly orchestrated slow-motion sequence in *Avengers: Age of Ultron*. We see all the Avengers in action at the same time against the swarming army of Ultron's robots, as part of the Battle of Sokovia in Novi Grad city.

It's worth taking a moment to mull over the backstory of that battle. Ultron, the allegedly all-wise artificial intelligence peacekeeping program devised by Tony Stark, decides that humanity is the greatest threat to peace on Earth. (It's a fair conclusion—you just have to consider the texture of this new century: a crumbling environment, nuclear stockpiles, unremitting government surveillance, the increasing possibility of rogue pathogens, and political leaders who seem to have walked straight out of a comic strip). Ultron decides to cut to the chase and help humanity commit global *seppuku*. His plan? Create a device that will lift a section of Novi Grad skyward, then send it swiftly earthward, causing it to create a meteoric crash when it impacts with the planet.

There are two chief problems with Ultron's lifting and crashing plan: first is the lifting, and second is the crashing. Let's deal with each in turn.

TAKING SOKOVIANS ON A SKY VACATION

It's quite common in science fiction movies for humans to be taken on a trip off-Earth, but it's not so common to try taking a bit of the Earth along for the ride. To work out how much energy Ultron needs to figure into his brainbox equations to perform this conjuring trick of lithic levitation, we first need to estimate the sheer mass of the lump of Sokovia he's trying to lift.

Taking another quick look at the movie footage; a fair estimate of the section of Sokovia that Ultron raises up is about two kilometers of the city, one kilometer each side of the church that sits at the epicenter of the rising rock. Once the city is airborne, it looks like a mini-Earth or, more appropriately, an asteroid. A quick reference to NASA's online guide of the mass of selected asteroids shows that a two-kilometer clump of rock and dirt, such as Geographos, has a mass of 0.004 x 1015 (or four trillion) kilograms. (By the way, Geographos appears to be a very good fit for our purposes. Named after the National Geographic Society, Geographos was discovered in 1951 by a team at the Palomar Observatory in California, and is one of the asteroids that sits in a potentially hazardous near-Earth orbit).

If Ultron were a plain old protocol droid, then he'd probably try to raise the rock using rockets. But Ultron ain't dumb, apparently. He knows that rockets also have to lift their own fuel. And so, to the four trillion kilograms of Sokovian rock he'd need to add a commensurate amount of fuel. True, he'd not need to calculate carrying that fuel all the way up, as the good thing about old-school rocket fuel is that, as it burns, the rocket gets lighter and lighter, which means the rock would need less and less fuel. Even so, to lift four trillion kilograms of Sokovian rock would require tens of trillions of tons of fuel. You can imagine Ultron might be sorely tempted with this scenario. After all, if those trillions of tons of fuel were hydrocarbon-based, it would represent a sizable percentage of the doomed world's remaining reserves, something that would surely amuse the batty bot.

But Ultron, we're told, is as brainy as bots can be, and has far more sophisticated tech up his metal sleeve. One of his options, perhaps, is the space elevator. A mainstay of science fiction, the space elevator idea is that if you connect a long enough cable to a space satellite in geo-stationary orbit, an elevator can be used to transport stuff into space using far less energy. No doubt Ultron feels raising Sokovia in this way simply wouldn't work. Besides, this option might lead to rather lame dialogue such as Ultron declaring, "Do you see? The beauty of it, the inevitability. You rise, only to fall. You, Avengers, you are my meteor, my swift and terrible elevator, and the Earth will be dented a little bit when I cut the cable. Mwahahaha, etc."

Let's also take this brief opportunity to help Ultron a little with his terminology. Rocks and debris in space are called meteoroids, and pieces that survive the journey to Earth and hit the ground are called meteorites. But the term meteor, and its colloquial equivalents "shooting star" or "falling star," is merely a reference to the visible incandescent passage of a glowing meteoroid falling through our planet's atmosphere on its way down to the ground. (Besides, something as big as this Sokovian mass of rock is surely an asteroid.)

Ultron would no doubt be aware that another lift option would be to blow Sokovia into the sky by using a nuclear weapon. This might appeal to him. The basic plan would be to place a nuclear bomb somewhere under the Sokovian city and simply ride the shockwave into the sky. You might assume that the city would be vaporized. And yet, someone with Ultron's alleged skill could no doubt fashion some kind of (probably) metal shield to stop the soon-to-be-floating city from disintegrating before it shot skyward.

Incidentally, Ultron could have stolen his lifting and crashing idea from the age-old proto-science fiction novel, *Gulliver's Travels*, written in 1726 by Jonathan Swift. Swift's book features a "flying island," made mainly of metal and measuring four and a half miles in diameter. This flying island called Laputa dominates the country above which it soars and, as with Ultron's plan, any discontent down below on the planet's surface can be destroyed completely by having Laputa itself plummet to Earth.

In the final analysis, it seems that Ultron's propulsion method of choice remains something of a mystery. There are screenplay references to him activating the "Vibranium core, and the Earth around Sokovia starts to shake and break." Otherwise, the screenplay is predictably vague about the lifting. But, let's assume the Vibranium core generates a force field, which somehow pulls the rock together as a superstructure so that it can be lifted without falling apart. Like Nick Fury's Helicarriers, it may be that the Vibranium also supplements the power of massive repulsor engines underneath the Sokovian bedrock. (Repulsorlift tech can be found in a number of places in science fiction, cropping up in the *Halo* video game universe, as well as in *Star Wars* and Marvel.)

Repulsorlift is usually seen as a technology that enables a craft to hover or even fly above the surface of a planet by pushing against gravity, resulting in the craft needing far less thrust. In the case of Sokovia, Ultron could be using a repulsorlift engine design that also secretes some kind of antigravity device. That way, the city could float without artifice, resisting the Earth's gravitational field. To raise Sokovia to a decent height above the Earth's surface, the mass of rock would need to apply an equal and opposite push to the gravitational force of the Earth (according to the principle of balanced forces in physics, first introduced into the scientific culture by British scholar Sir Isaac Newton, of course). But for acceleration in an upward direction from the Earth's surface, Ultron would also need his repulsorlift engines to apply a thrust.

Now, according to Einstein, gravity is just curved space, so, all Ultron needs to do to create antigravity is simply bend space the other way. If the geometry of space can be bent to his will, then he can create antigravity, with the ability not only to float objects freely, but also even raise huge rocks into the sky. All mass makes gravity, but it won't be easy for Ultron to create a material that makes antigravity. He could possibly use exotic matter. Theoretically, exotic matter has negative mass, which should create the reverse effect of gravity and be used to cancel out the weight of a city, just like any other repulsorlift craft.

Assuming Ultron is an engineering wizard, by no means certain on the evidence so far, he'd have to work out just how much exotic matter he needs to add to the mix. He'd have to estimate the chunk of Sokovia correctly at four trillion kilograms, and ram into the rock an equal mass of the exotic stuff. Presto, the resultant mass of Sokovia is zero.

And yet two outstanding problems remain for Ultron in terms of the lifting aspect of his pet project. First is the sourcing of exotic matter, since no one at the moment is really sure what it actually is. The second is that, given everything else he seems to have miscalculated, there is still some doubt that Ultron would get his repulsorlift calculations correct. And that brings us to the other aspect of his project—Ultron's crashing plan.

MAKING AN ASTEROID OF SOKOVIA

Let's just recap Ultron's cunning crashing plan. The lifting bit notwithstanding, Ultron's plan to wrestle robot deliverance is (with a philosophical nod to the demise of the dinosaurs) to use Sokovia to create a similar extinction-level event to end humanity. Once raised to a sufficient height,

Sokovia can then come crashing down in a calamitous impact with the Earth.

But Ultron will have to work out just how much energy is at his disposal, assuming he's got his calculations right about the mass of Sokovia being four trillion kilograms. The secret to Ultron's success is speed. The faster the city plummets, the closer it is to curtains for the human race. And to get that speed of extinction, he'd have to let the city drop from a huge height.

It's all down to energy, potential, and kinetics. When Sokovia is raised to a height above the Earth, like all bodies in a gravity field, it gains gravitational potential energy. When Ultron's grip on the rock city is released, this potential energy is converted into kinetic energy, as the now free-falling Sokovia plunges down to the Earth below.

How much of a height above the Earth's surface is enough for Ultron's crashing plan? In the movie, the city is raised to such a height that Captain America remarks that "the air is getting a little thin." Assuming this is the height where altitude sickness first sets in, that would mean the rock was about 2.5 kilometers up when its descent began. That's nowhere near enough.

If you ignore air resistance—which we really shouldn't, but let's make this as simple as we can—a free fall from this height would deliver a fifty-megaton impact to the very heart of Sokovia. That's the same impact as the most powerful nuclear weapon ever detonated on Earth, Russia's "King of Bombs," tested on October 30, 1961 and the most powerful man-made explosion in history. But, as we're all still here and I'm sitting writing this, a fifty-megaton impact is clearly not an extinction-level event—not unless reality is an illusion and I'm actually in *The Matrix*. So, in all probability, it's back to the drawing board for Ultron.

Assuming the threshold to turn all *homo sapiens* into *homo extinctus* (spoiler: not an actual species, yet) is around 100,000 megatons, Ultron would either have to plump for a bigger clump of Sokovia or raise his already established rock to a height of around 10,000 kilometers above our planet. He's ticked the box of having the correct gravitational potential energy, and hits the jackpot, but for a true dinosaur-level extinction event and human wipeout, this drop height of 10,000 kilometers or 6,300 miles would have to be increased thousand-fold to 6.3 million miles above the surface.

Whichever option he picks, Ultron's got problems. If he goes for the full dinosaur destroyer option of 6.3 million miles, the Earth's gravity pull is far less at that distance, which means his "meteor" wouldn't have the impact energy it needs. And if he goes for the 6,300 miles, he'd be faced with the prospect of having to use thrusters to turbo boost Sokovia on its asteroidal approach to Earth, as the city would have to be speeding at 14 kilometers a second at impact.

Now some might say that the true hero of the Battle of Sokovia is Jarvis, who managed to deflect Ultron away from the more sensible global kamikaze targets of our nuclear arsenals and power plants and our water supplies. But given Ultron's record on his lifting and crashing plan, he'd no doubt make a mess of any plan B he might have, too.

DAILY DIARY: IRON MAN, SUPERMAN, AND COPING WITH THE PHYSICS OF FLIGHT

"The human bird shall take his first flight, filling the world with amazement, all writings with his fame, and bringing eternal glory to the nest whence he sprang."

—Leonardo da Vinci, *Codex on the Flight of Birds* (1505)

"For once you have tasted flight you will forever walk the Earth with your eyes turned skyward, for there you have been, and there you will always long to return."

—Leonardo da Vinci, *Codex on the Flight of Birds* (1505)

"All in all, for someone who was immersed in, fascinated by, and dedicated to flight, I was disappointed by the wrinkle in history that had brought me along one generation late. I had missed all the great times and adventures in flight."

—Neil Armstrong, *Neil Armstrong: Quotes and Facts* (2015)

Modern daily life can be a drag. Sure, science and tech have brought us the wheel, New York, the smartphone, etc. But dwell on the drawbacks a little: spam emails, self-serve tills, and predictive text, to name but three. Then there's snail's-pace wi-fi, being a sociopath on social media, and dealing with trolls on Twitter.

But worst of all is the terrible traffic. There can be no greater need for the dream of human flight than today's jam-packed world of traffic. Take China, for example. September 2013 saw the world's longest traffic jam, more than 100 kilometers long and lasting for weeks. The problem is so common in China that some folk have embraced it as an entrepreneurial opportunity. Their motorbike businesses will weave their way between the gridlocked lanes and take you to your destination. Hell, they'll even provide someone to sit in your car for you until the jam is done, if it ever ends. Pizza deliveries to jammed cars are also very common. Pizza express, even if the traffic isn't.

Little wonder flight is so popular among entrepreneurs. In 2015, American business magazine *Forbes* found that flight was the business leaders' superpower of choice. Collecting data from professional leaders who read their blogs, *Forbes* posed the question, "If you were given a choice of two special powers, which would you prefer? A. Ability to fly, or B. Power to be invisible."

Flight came first in all classes. *Forbes* gathered data from 7,065 leaders around the globe: 63 percent of the data from North America, 13 percent from Europe, 16 percent from Asia, with 8 percent from all other responding territories. Clocking up superiority over invisibility of almost three to one, 72 percent of business leaders chose the ability to fly over being invisible. Data analysis of the corporate positions of responders found that 76 percent of top managers selected the ability to fly, as compared to only 71 percent of individual contributors. The idea of human flight is pretty popular.

But if humans did know natural flight, just how fast would we fly? The first human to break the speed of sound, mostly without using tools or tech, was Austrian skydiver Felix Baumgartner. This daredevil and BASE jumper reached a maximum velocity of 833.9 mph (1,342 km/h) on October 14, 2012. Felix launched himself out of a balloon at 128,100 feet (that's 24 miles or 39 kilometers) above New Mexico.

Felix almost failed the dive, as his helmet visor fogged up. His twenty-four-mile flight took just under ten minutes, with the last few thousand feet of the descent done by parachute. Felix was quoted by the BBC after the flight as saying, "Let me tell you—when I was standing there on top of the world, you become so humble. You don't think about breaking records anymore, you don't think about gaining scientific data—the only thing that you want is to come back alive."

But Felix's feat wasn't flight, it was freefall. Now, consider footspeed. Whereas forty-three-year-old Felix smashed the record for the highest ever freefall, twenty-two-year-old Usain Bolt broke the footspeed record. Footspeed, also known as sprint speed, is the maximum speed at which a human can run. Bolt's record was recorded at 27.8 mph (44.72 km/h) during the 100-meter sprint final of the World Championships in Berlin on August 16, 2009, five days before Bolt's twenty-third birthday. The average speed Bolt clocked over the course of the race was 23.35 mph or (37.58 km/h). But Bolt owes his footspeed to Newton. Usain's sprint speed depends on how much force is brought to bear by Bolt's formidable legs, and according to Newton's Second Law of Motion, that force is the product of mass and acceleration. Newton's Third Law says that for every action, there is an equal and opposite reaction. In the case of the mechanics of footspeed, this translates into Bolt's running action needing a firm ground to push against, with the ground effectively pushing back against Bolt.

MOVING SWIFTLY THROUGH WATER AND AIR

Flight is far more similar to swimming. You've no doubt seen the frantic freestyle swimmers stroking swiftly through their lanes at the Olympic Games, but are those competitors really chopping through the water like sharks? Nope. Truth is, those guys are hardly moving. The fastest recorded human swim speed is less than five miles an hour. Hell, a toddler torpedoed by a self-inflicted tantrum can outrun those Olympic swimmers.

And the reason why goes back once more to Newton. As Bolt runs, he speeds along because his legs push against the track with his feet and the track pushes back, thrusting him forward. Athletic tracks are solid. And that means the particles in the track are basically bonded together and must push back against Bolt, rather than simply moving out of the way. The same goes even for the torpedoed toddler: firm ground. But the Olympic swimmers have another medium to mess with. Water is fluid and flows far more easily. When those Olympic leviathans plunge their limbs to push back against the water, some of the water molecules are easily able to slip past one another rather than pushing back against the swimmer's limbs.

Now, think about applying our swimming lesson to the question of flight. Air, like water, is also a fluid, but the gas particles that make up air are far freer to move about than the liquid molecules of water. As gases are less dense than liquids, air has more free space for particles to slip and slide past one another, so a human flyer would waste more energy than a human swimmer because a lot more air would have to be pushed backward to be able to move forward.

Consider Sandra Bullock as Dr. Ryan Stone in the 2013 movie, *Gravity*. Stone is an astronaut stranded in space after the mid-orbit destruction of her space shuttle, and is trying against all odds to return to planet Earth. In some scenes we see Stone moving around a spacecraft in microgravity. How does she do it? She doesn't waste time flapping around in the near vacuum of space. She simply pulls on handles installed on the ceiling, walls, and floors of the craft to get purchase and make headway.

Imagine, like Stone, you were able to float. Not in the microgravity medium of space, but down here on good old Earth. Exactly how would you get purchase to move about from block to block in the middle of the street? You couldn't do a Spiderman and swing from a web. And swimming through the air wouldn't get you very far either, but let's assume we make less fuss of this physics and grant you the fishy ability to float. Maybe some kind of antigravity thing is going on. Yeah, that's it. And let's also assume you are free to speed about using some kind of thruster tech previously unknown to man.

STRAIGHTEN UP AND FLY RIGHT

How high would you be able to go? Remember that scene when Iron Man zooms up through the lower atmosphere, taking his suit on its first flight, and he finds it has a problem with freezing at high altitudes? Well, there's an area of physics around the behavior of gases, and it's known as thermodynamics. One of the laws of thermodynamics, the Ideal Gas Law, says that the pressure and temperature of a gas increase and decrease together. Written down in an equation, the Ideal Gas Law looks like this: $PV=nRT$, where P is the pressure of the gas, V is its volume, and T its temperature. N is the quantity of gas molecules in moles, and R is the ideal gas constant, 8.314 J/Kmol.

As you fly up into the atmosphere, there is less pressure. Air expands in volume with less pressure, so the molecules have more room to wander around without colliding into each other and creating heat. And as the pressure is a lot lower at high altitudes, it would be freezing cold if you were flying above the clouds, so you would need to keep your core body temperature warm or you'd soon begin to shiver violently, become mentally muddled, and plummet out of the sky due to lack of muscle control from hypothermia.

Volume would be a problem, too. The Idea Gas Law shows that as pressure decreases, gas volume increases, so if you were to fly straight up too quickly, the gas inside your body would expand rapidly. It's like when soda fizzes up when jiggled. The name for this is decompression sickness, which is perhaps better known as "the bends," and it's the phenomenon associated with the experience deep-sea divers get when they come up too quickly. It's been known since 1670, when Irish chemist, Robert Boyle, experimenting with a viper in a vacuum, showed that a reduction in ambient pressure could lead to the formation of a bubble in living tissue. In humans too, the bends results in pain, paralysis, or death, based on how foamy your blood becomes.

So, let's say we stray away from flying too high. If we keep our flight paths a little closer to the ground, what then? On the plus side, we'll be able to see all the road furniture, mostly meant for regular traffic. There's also the added advantage of being able to breathe oxygen with ease! Best copy winged superhero, Falcon, though, and goggle up. You'll probably need a helmet like Captain America, too, for protection against electrical cables, insects, birds, and high-hanging street signs.

On the more dangerous side, there are the drones. Sure, there are also other flying humans, including flying cops ready at the drop of a wing to gift you a ticket for jay-flying in the wrong intersection of air or something, but that's nothing compared to the coming threat of drones. No doubt some day in the future NASA and others will have finally nailed drone traffic management. But until that day, beware of the drones.

For the skies are getting crowded. The Federal Aviation Administration (FAA) of the United States and its European counterpart the EASA report that the number of near-misses with drones has surged since 2014. There were as many as 650 cases as of August 2016. Dubai airport has been repeatedly shut down by drone activity, and in July 2017 a pilot reported to the Australian Transport Safety Bureau that his light aircraft was struck by a drone ahead of landing in Adelaide. In time, these drone mechanical birds may potentially grow bigger and more dangerous.

But maybe Leonardo da Vinci is right. Once you've tasted flight, you will always long to return skyward. Sure, you may have a drone collision in mid-air which knocks you senseless so you find yourself in free fall, just like Felix Baumgartner. But, ignoring aviation authorities and some of the laws of physics, most potential flyers would opt for flight, as they gaze down at the miles-long traffic jams below.

WHY MIGHT STORM'S WEATHER-WIELDING BE UNWISE?

Exterior. Wooded hillside.

A sudden flash of light.

The wind blows so violently now that Sabretooth nearly misses two figures standing only a few yards away—mere silhouettes in the icy haze. A closer look tells us it is a man and a woman, they wear strange uniforms of form-fitting material—the woman's face is bare, revealing dark skin and penetrating eyes.

Storm looks down . . . concentrating her intense gaze. The wind whips further, and her skin dimples with goosebumps as the temperature drops and the water . . . begins to freeze. Storm then further forces the temperature down, freezing the ice thick and thicker . . .

The snow and wind are now violently raging.

—Ed Solomon and Chris McQuarrie, *X-Men* early screenplay draft (1999)

"It used to be thought that the events that changed the world were things like big bombs, maniac politicians, huge earthquakes, or vast population movements, but it has now been realized that this is a very old-fashioned view held by people totally out of touch with modern thought. The things that really change the world, according to Chaos theory, are the tiny things. A butterfly flaps its wings in the Amazonian jungle, and subsequently a storm ravages half of Europe."

—Neil Gaiman, *Good Omens: The Nice and Accurate Prophecies of Agnes Nutter, Witch* (2006)

"We are all connected; To each other, biologically. To the earth, chemically. To the rest of the universe atomically."

—Neil deGrasse Tyson, *Symphony of Science* video, *We Are All Connected*, YouTube (2009)

"You've never heard of Chaos theory? Non-linear equations? Strange attractors? Living systems are never in equilibrium. They are inherently unstable. They may seem stable, but they're not. Everything is moving and changing. In a sense, everything is on the edge of collapse . . . God creates dinosaurs. God destroys dinosaurs. God creates man. Man destroys God. Man creates dinosaurs. Dinosaurs eat man. Woman inherits the earth."
—Michael Crichton, *Jurassic Park* (1990)

Ever since the first irresistible rise of science, the mission was never just to explore nature, but to exploit it. The machines of science were created with nature's dominion in mind, and science fiction took that mission into a far-flung and imagined future. The explorer who seeks to penetrate space as in Jules Verne's *Journey to the Center of the Earth* wishes to possess nature absolutely for science. To reach the core of the world is to achieve completion, to pierce the living heart of nature, the glittering prize. Likewise, in H. G. Wells's *The Time Machine*, the time traveler sets out to navigate and dominate time. His doom-laden discovery for science and for the human race? Time is lord of all. The significance of the story's title becomes clear: man is trapped by the mechanism of time, and bound by a history that leads to his inevitable extinction.

And high on the list of the challenges facing science in its quest to tame nature sits the weather: how to master the chance and capricious whim of forces that can't even be seen, let alone predicted. Most people find that the challenge of forecasting the weather is that the forecast is right too often for us to ignore it, and wrong too often for us to rely on it. Weather sometimes seems to be one of nature's greatest secrets mocking humans. And yet the abiding rain has soft sculpting hands with the power to cut stones and chisel sheer drama into the shapes of the very mountains while creating rainbows in an apparent apology for angry skies. Little wonder that scholars have long dreamt of controlling the weather. The right amount of rain and sun means healthy crops, shelter, and prosperity; too much or too little, hunger and death.

In some ways the creation of a fictional superhero who could master the weather was inevitable. Ororo Munroe first appeared in 1975 and is most commonly associated with the X-Men. Storm's backstory is a tale of hereditary mutation. Her mother was a Kenyan princess from a long line of African witch-priestesses with signature blue eyes and white hair. The witch lineage is a line of mutants born with the superhuman ability to control the weather. On the untimely death of her parents, Storm ends up being worshiped as a goddess once her powers manifest, but is eventually recruited by Professor X. Storm soon becomes a member of the X-Men, leading them from time to time, while also working with the Avengers and the Fantastic Four.

WHETHER WIELDING THE WEATHER IS WISE

And yet, might Storm's weather-wielding be rather unwise? Given our current problems with climate change, is meddling with the weather to be admired? For example, what about the so-called Butterfly Effect? This idea first found its voice in science fiction with Ray Bradbury's moral fable, *A Sound of Thunder*, written in 1952. In Bradbury's tale, a time-tourist wreaks temporal havoc by treading on a prehistoric butterfly and unleashing an alternate timeline. The story told of sensitive dependence upon initial conditions. It was written a full ten years before early pioneer of chaos theory Edward Lorenz developed its principles for the scientific community through mathematics and meteorology. Since then, writers like Neil Gaiman have written about how "a butterfly flaps its wings in the Amazonian jungle, and subsequently a storm ravages half of Europe." Michael Crichton in *Jurassic Park* tells the dramatic tale of what happens when living systems, inherently unstable and forever moving and changing, are pushed over the edge and collapse.

Let's conjure up our own Storm weather-wielding scenario, and work through the possible consequences of her actions. For the sake of argument, let's say Storm is battling with Magneto. Hardly a stretch of the imagination for, as good and bad guys go, Magneto rarely seems to know what side he's really on. And since he's one of the most powerful and dangerous members of the Brotherhood of Evil Mutants, he's more than a match for Storm with his forceful ways, manipulation of electromagnetism, and sheer will to do something insane—like shut down the Internet or some other despicable atrocity.

The battle doesn't start well for Storm, and Magneto quickly gets the upper hand, so Storm comes to an epic conclusion: she decides to fight fire with fire, or at least electromagnetism with electromagnetism. She plans to focus all her spooky Kenyan witch power and aims to telepath the sum of all lightning strikes currently occurring around the globe. Her noble goal is to finally zap Magneto's ass, once and for all. Storm knows there's power in a lightning strike and, as lightning is electricity, it should have a decent zap factor—as long as she can get all the lightning to strike where Magneto stands.

Storm sets her witch-priestess thoughts in train. She knows a typical lightning strike harbors enough energy to power a typical home for a couple of days. She also knows that even in places on the Earth's surface that see a lot of lightning such as the Congo, the power delivered to the ground by lightning is a million times less than the power brought by sunlight. But Storm's master plan, or mistress plan perhaps, is to conjure up *all* the lightning. To make a mega-bolt, in which all the lightning comes down in parallel, bunched up in one big bolt.

Storm is aware of the fact that the main channel of a lightning bolt—the zap that carries the current—is around one centimeter across, so Storm's mega-bolt, the sum of all strikes and containing around a million discrete bolts, will measure around six meters in diameter. More than enough to zap Magneto where he hovers. Storm thinks about the way that Magneto is often boasting about his powers in measures of Hiroshimas: "Wow, did you see that, Storm, Niagara Falls may deliver the power of a Hiroshima every eight hours, but I just did it in a single second," or "The soft breeze that blows across a prairie might carry the kinetic energy of a Hiroshima, but, Storm? I just made everyone's hair stand on end in midtown Manhattan." That kind of thing. But now, Storm is about to get her own back. The mega-bolt coming Magneto's way will zap him with *two* atom bombs' worth of energy.

Somehow (this is superhero stuff, after all), Magneto divines Storm's intent. He quickly does some mental calculations of his own. These are mostly focused on the ideas of: Plan A, getting his ass out of the way, or Plan B, somehow dissipating the power that is about to fall from heaven above. Totally out of character, Magneto doesn't think of the incoming energy in terms of Hiroshimas; maybe he's in denial, or simply doesn't want to credit Storm with such superhero proportions. Instead, he figures there's enough electrical energy in the imminent mega-bolt to power a games console and plasma television for many millions of years, or to feed America's hungry domestic need for electricity for just five minutes.

Magneto mulls his options. The girth of the mega-bolt would measure about the same as the center-circle of a basketball court, but the damage it would do would wreck the very court itself. Not much room for Magneto to maneuver. Besides, inside the mega-bolt, the air would be transformed into high-energy plasma, and that means the light and heat from the bolt would spontaneously ignite surfaces for miles in all directions. The very shockwave of the mega-bolt would fell trees, and flatten buildings, leaving Magneto with a dwindling list of choices.

So instead he considers his Plan B: Rather than merely sidestepping the mega-bolt, somehow managing to deflect it. It would easily be within Magneto's powers to draw some kind of lightning rod toward him, and try to ward off the mega-bolt that way. But it's not the best of his options. For one thing, no one really knows how lightning rods work. One theory is that they ward off bolt power by "earthing" charge from the ground into the atmosphere, thereby lowering the difference in electrical potential between the clouds and the ground. This is thought to reduce the probability of a strike, but since Storm is doing this on purpose, a rod is hardly likely to help him in his current predicament. Say, for example, that Magneto was able to use his powers of affinity to summon a copper cable a meter thick and hovering just touching the ground. Sure, the brief torrent of current from the mega-bolt would be conducted by the copper without melting it, but when the bolt shot down to the bottom of the rod and made contact with the ground, there would be an explosion of molten rock to deal with, and Magneto was less happy with lava than he was with lightning.

But then the coin drops in Magneto's mind: suddenly he remembers Storm is deeply claustrophobic. Magneto recalls from his knowledge of X-Men history that, although she is a very strong woman, Storm, like Superman, has one main weakness—her claustrophobia. The phobia dates back to her childhood in Cairo, when a jet crashed into her home, killing her parents. The impact brought the house down on young Ororo, burying her alongside her dead parents under the rubble, and forcing her to wait days for rescue, lying next to their corpses. The incident left Storm with a severe phobia of enclosed spaces, a phobia that Magneto could now exploit. Not for nothing is he a member of the Brotherhood of Evil Mutants. Having so much experience with lifting objects such as missiles, guns, cars, and even the entire Golden Gate bridge, Magneto is easily able to create a made-to-measure metal cage in which to fence and befuddle the witch-priestess. And that's the thing with the non-linear equations of chaos theory. Storm messes with Magneto. Storm wields the weather. Magneto messes with metal. Storm catches a new strain of claustrophobia. You can cage the wielder, but not the weather.

WHY ARE SUPERVILLAINS ALWAYS INVADING EARTH?

"No one would have believed in the last years of the nineteenth century that this world was being watched keenly and closely by intelligences greater than man's and yet as mortal as his own . . . With infinite complacency men went to and fro over this globe about their little affairs, serene in their assurance of their empire over matter . . . At most, terrestrial men fancied there might be other men upon Mars, perhaps inferior to themselves and ready to welcome a missionary enterprise. Yet across the gulf of space, minds that are to our minds as ours are to the beasts that perish, intellects vast and cool and unsympathetic, regarded this Earth with envious eyes, and slowly and surely drew their plans against us."

—H. G. Wells, *The War of the Worlds* (1898)

So begins H. G. Wells's novel, *The War of the Worlds*, arguably the most superb opening to a story in the entire history of science fiction. *The War of the Worlds* features the first ever "menace from space." Mars is a dying world. Its seas are evaporating, its atmosphere dispersing. The entire planet is cooling, so "to carry warfare sunward is, indeed, their only escape from the destruction that generation after generation creeps upon them." So the terror of the void is brought to Earth. In his tale, Wells issues repeated reminders of "the immensity of vacancy in which the dust of the material universe swims" and invokes the "unfathomable darkness" of space. Life is portrayed as precious and frail in a cosmos that is largely deserted. Wells's book carefully conveys the quality of the void—immenseness, coldness, and indifference—in its rendering of the aliens. And his Martian machines vividly hammer home the cosmic chain of command, "It is remarkable that the long leverages of their machines are in most cases actuated by a sort of sham musculature . . . Such quasi-muscles abounded in the crablike handling-machine . . . It seemed infinitely more alive than the actual

Martians lying beyond it in the sunset light, panting, stirring ineffectual tentacles, and moving feebly after their vast journey across space."

The modern alien owes everything to Wells. Wells's Martians are agents of the void. Their distinctive physiology and intellect made them the prototypical alien. Their Tripods tower over men physically, as the vast intellects of their occupants tower over human intelligence. Bodily frail, but mentally intense, the Martians and their superior machines are instruments of human oppression. Their weapons of heat rays and poison gas are dehumanizing devices of mass murder. All attempts at contact are futile, furthering the idea of the aliens as an unrelenting force of the void.

And nowadays, it seems, supervillain aliens are forever invading our humble little planet.

Most famous is the Battle of New York. Also known locally as "the Incident," the Battle of New York was a week-long clash of titans, with the Earth and the Avengers on one side, and Loki and the Chitauri army on the other. But that's just the latest example. There's also the subversive, long-term secret invasion of Earth by the shape-shifting alien Skrulls, a story that ran in Marvel comics April through December 2008.

In fact, alien invasion has become an inexhaustible topic for fantastic film and fiction. But the sophistication of alien depiction has developed little since *The War of the Worlds*. Wells's novel says nothing about Martian culture. It seems to have wasted away by some entropic decay and whittled down to nothing more than a cosmic justification to invade Earth. The Martians have no interest in human culture. Like vampires, they are interested only in human blood, and this clinical rationale for human oppression inspires the readers, loathing of the Martians, and the latent power of unsolicited natural selection.

Wells's Martian invasion is justified. The vast majority of supervillain invasions are not. Wells's Martians are inhabitants of a dying planet which is fast winding down into desert. But this exceptional case for alien invasion within our solar system has been thoughtlessly adopted for a whole genre. Writers of film, fiction, and comic books have mostly aped the Wellsian invasion myth, but without the exceptional case Wells made for the Martians. Not only that, but in an attempt to eclipse the master and his Martians, writers attributed ever greater power to the aliens. And they gifted to Earth the promise of unimaginable riches, a glittering prize of a planet not only of value to the small desert world of Mars, but for any imaginable civilization in the Galaxy, and beyond!

KRYPTON CALLING

Take Superman, for example. While mostly not a supervillain as such, The Man of Steel was famously born as Kal-El on the alien planet of Krypton. His parents, Jor-El and Lara, learn of Krypton's imminent destruction, sounding exactly like the kind of entropic decay H. G. Wells envisioned for Mars. So, Jor-El starts to build a spacecraft to carry Kal-El to Earth. Now, according to *Superman #132*, planet Krypton was three million light years away from our world. Cosmically speaking, that's not so far, but it's hardly our next-door neighbor either. Over twenty other galaxies sit between Superman's world and the Milky Way.

What, we might wonder, inspired Jor-El to send his son Earthward? Let alone the exacting logistics of launching a rocket on a light years' trajectory in this direction. Given there are certainly billions of planets between Krypton and Earth, and assuming there are probably thousands of civilizations per galaxy, you get a good idea of the lethargic thinking behind the plan.

Like Wells's *WOTW*, the Superman story has no culture clash between earthlings and aliens. Clark Kent is raised in secret. There is no great revelation that he's an alien from a sophisticated and advanced civilization, which is perhaps just as well as it would make the reader wonder why Jor-El's race of supermen had even bothered with our humble little world. Astronomers recently estimated there are a dizzying two trillion galaxies in the Universe. That's up to twenty times more than was previously thought. And this recent revelation was based on 3D modeling images collected over two decades by the Hubble Space Telescope. Even if there are only a thousand civilizations per galaxy, that puts the number of advanced civilizations in the cosmos at two thousand trillion. With all this real estate in the Universe, Earth would have to be very exceptional to justify a visit.

Things look even more incredulous when you consider the invasions of supervillain aliens. Powerful armies, with vast and monstrous masses of starships at their disposal, are all hell-bent on taking over our tiny world. Technology has sharpened their cosmic fangs, all the better to eat humanity. And yet the difference in scale is like assuming an earthly superpower, such as China, is dead set on mobilizing its armies to expropriate the local corner store. In reality, the readers know that the cost of invasion must be worth more than the value of the booty.

THE SUPERVILLAIN COSMOS

And yet you never really know with this supervillain cosmos. Maybe the aliens simply aren't motivated by material gain. Perhaps these supervillain aliens attack our planet merely because it pleases them to do so. They destroy for the sake of destruction. They enslave humanity as an amused and academic exercise in despotic mastery. Shame, really. It's a far cry from Wells. *The War of the Worlds* had been an exercise in interplanetary Darwinism. Wells's imaginative lens had been like a type of telescope; the invading Martians are the "men" of the future. But it's the wrong end of the telescope—Imperial Britain is on the receiving end of social Darwinism.

Wells's wrath is focused on the idea of the becoming of man—the prevailing idea at the time that polite English, middle-class society was the very point of evolution. The target of Wells's parody was the same kind of flabby thinking that Mark Twain had superbly satirized in his wonderful essay, "Was the World Made for Man?" where Twain says "Man has been here 32,000 years. That it took a hundred million years to prepare the world for him is proof that that is what it was done for. I suppose it is. I dunno. If the Eiffel Tower were now representing the world's age, the skin of paint on the pinnacle-knob at its summit would represent man's share of that age; and anybody would perceive that that skin was what the tower was built for. I reckon they would. I dunno."

Likewise, Wells's *WOTW* begins with a quote from German genius Johannes Kepler: "But who shall dwell in these worlds if they be inhabited? Are we, or they, Lords of the World? And how are all things made for man?" The narrator of Wells's story of a struggle for survival is a philosopher, writing a thesis on the progression of moral ideas with civilization.

His conclusion of a bright future is abruptly blown apart in mid-sentence by the brutal natural force of evolution in the shape of the Martian attack.

Wells had planned his Martian invasion on bicycle. It is pleasing to picture him mapping mayhem as he "wheeled about the district marking down suitable places and people for destruction by my Martians." As early as 1896 he declared his intentions to "completely destroy Woking—killing my neighbors in painful and eccentric ways—then proceed via Kingston and Richmond to London, which I sack, selecting South Kensington for feats of particular atrocity." It is this exquisite violence of Wells's imagination that marks his genius.

The vast majority of alien invasion stories since show little evidence of planning, imagination, or genius. Wellsian interplanetary Darwinism seems to have been replaced by a kind of paranoid philosophy penned by the Marquis de Sade. This imagined paranoid cosmos seems obsessed with the conquest of (mostly harmless) Earth. It is a cosmos that sets every type of trap to catch humankind, whether by outright attack, or by stealth and robbing us of our free will (a scenario that proved very popular during the days of the Cold War and the Senator Joseph McCarthy years).

The paranoids and hypervigilants are with us still. They apply the same logic to the SETI or METI debate about whether or not it's okay to send messages to nearby stars in the hope of attracting the attention of their hypothetical inhabitants. In short, whether to beam yoo-hoo messages to ET. No doubt having read the repeated and monotonous alien invasion scenarios borrowed from Wells but now showing a poverty of philosophy, the paranoids fear aliens will invade because (a) they sadistically feel like it, (b) they fancy a cosmic game of cops and robbers, or (c) they are a kind of intergalactic band of Jehovah's Witnesses or Scientologists come to save us from ourselves.

When it comes to invading super-aliens, it's high time for a change. Humans are no angels, and would surely have no qualms about killing a cockroach, but they would hardly go to the ends of the Earth to do so. Likewise, aliens, if you'll forgive the comparison of humans with cockroaches. Such super and sophisticated civilizations would have no need to go out of their way seeking earthlings to swat.

WHAT IF WE COULD USE SUNLIGHT TO POWER OUR BODIES LIKE SUPERMAN?

"Earth's sun is younger and brighter than Krypton's was. Your cells have drunk in its radiation, strengthening your muscles, your skin, your senses. Earth's gravity is weaker, yet its atmosphere is more nourishing. You've grown stronger here than I ever could have imagined. The only way to know how strong, is to keep testing your limits."

—Jor-El, *Man of Steel* (2013)

"By blending water and minerals from below with sunlight and CO_2 from above, green plants link the earth to the sky. We tend to believe that plants grow out of the soil, but in fact most of their substance comes from the air. The bulk of the cellulose and the other organic compounds produced through photosynthesis consists of heavy carbon and oxygen atoms, which plants take directly from the air in the form of CO_2. Thus, the weight of a wooden log comes almost entirely from the air. When we burn a log in a fireplace, oxygen and carbon combine once more into CO_2, and in the light and heat of the fire we recover part of the solar energy that went into making the wood."

—Fritjof Capra, *The Web of Life: A New Understanding of Living Systems* (1997)

"Nature has put itself the problem how to catch in flight light streaming to the earth and to store the most elusive of all powers in rigid form. To achieve this aim, it has covered the crust of earth with organisms, which in their life processes absorb the light of the sun and use this power to produce a continuously accumulating chemical difference. The plants take in one form of power, light; and produce another power, chemical difference."

—Robert Mayer, in pamphlet, *The Organic Motion in its Relation to Metabolism* (1845)

"When you think about the complexity of our natural world—plants using quantum mechanics for photosynthesis, for example—a smartphone begins to look like a pretty dumb object."

--Jeff VanderMeer, on *BuzzFeed* (2014)

Superman simply isn't what he used to be. Over the decades of the evolving character of the Man of Steel, his powers and abilities have waxed and waned, as creators and writers have toyed with their creation. And yet there is one constant: he's a sunshine Superman. He gets the majority of his powers from the Sun: strength, speed, flight, invulnerability, as well as the more refined abilities of x-ray vision, heat vision, and freezing breath. Earth's yellow star essentially transforms Superman into a huge solar battery, so much so that he has the added ability of super flare, an omnidirectional heat blast that annihilates anything within a quarter mile radius, and which uses and expels all his stored up solar energy in one giant blast.

Canon has it that the powers of Superman depend on his cells' capacity to absorb and metabolize energy from yellow stars like the Sun. The return from hotter, blue stars would be even greater. How does this affect Superman's Kryptonian body? Simply in the way that his living "solar battery" metabolism absorbs sunlight and converts it into fuel, which then enables his superhuman skills. His body can also store sun energy, like a kind of organic capacitor, so Superman can defer the use of his powers to times when he might find himself in dark places, such as at night, or in space. The degree of these powers would be impossible on Krypton, as its sun, Rao, is a red supergiant. But to what extent does canon fit the facts? To answer that question, let's consider sunlight first, and photosynthesis second.

SUNSHINE FOR SUPERMAN

Firstly, sunshine is the source of all life on Earth. Without sunshine there would be no heat, no food, no weather, no light, no day or night, and no seasons. In fact, no Earth as we know it. The Sun acts as a furnace for all those things. And even though it's a million miles across, the Sun is still considered something of a dwarf by cosmic standards. Even so, it's fat enough to fit over one hundred Earths across its belly, and over a million Earths inside it. Just to fly around the Sun once—even if you're Superman taking it easy at a mere supersonic speed—would take 227 days.

Stars, like the Sun, burn differently from things on Earth. Stars use nuclear energy. And since stars are made mostly of hydrogen gas, the atoms of hydrogen are fused together in their interiors to make new atoms, which release lots of energy. And given the size of the Sun and the fact that it's so packed with stuff, the temperature at its very center is about sixteen million degrees centigrade—that's hot enough to burn or fuse the hydrogen into helium gas. All this means that the Sun burns about four million tons of gas every second. That's as much energy as seven trillion nuclear explosions every second.

But what exactly is this sunlight, which Superman's body is exploiting? It might seem like an odd place to start, but think about Superman in a swimming pool. When he dive-bombs into the pool (and gets shouted at by a very brave lifeguard) the energy created by his body hitting the water can be seen in all the waves, which move across the water. Note that the water wave is not made up of water. The wave is made up of energy. And it's the wave that moves, not the water itself.

Sunlight is also energy that travels in waves, but it can move through space and air as well as water. You can think of light as "packets" of energy called photons, which our eyes are able to see. There are lots of different ways to make photons of light, but in space starlight is made at the fiery center of stars like the Sun, burning up hydrogen gas.

As our Sun is a star, sunlight is the same as starlight, and starlight is the fastest thing in the universe. Light travels at a speed of about 300,000 meters every second or 186,000 miles a second. Light is so swift it can go around the Earth's equator ten times every second. It takes just one second to get to the Moon, and around eight minutes to get from the Sun to Superman himself.

So does it make sense that it is the high intensity of the sun's yellow radiation that gives Kryptonians their superpowers? Canon says that our sun has more energy than Krypton's red supergiant sun, Rao. Red supergiants certainly exist. To take a famous example, consider the constellation of Orion, in which there is a red supergiant star by the name of Betelgeuse. Betelgeuse shall be our guide to gauging what Rao might actually be like. If you could pick Betelgeuse up and plop it right where the Sun sits at the very center of our solar system, Betelgeuse's belly would be big enough to swallow up the orbits of the four innermost planets—Mercury, Venus, Earth, and Mars. That's some monster star.

The key to each star's temperature is its color. If Superman were to gaze up at the night sky, at first sight it would seem that all the stars were white. But, when he looked more watchfully, he would be able to spy that some stars are reddish, such as Betelgeuse and Rao; some are yellowish, such as our sun; and some are bluish, such as Rigel, another star in Orion. That's because a star's color is a tell-tale sign of how hot or cold it is. The blue stars are swelteringly hot. The red stars are a lot cooler, but still burning away at their very core. And yellow stars like the Sun are in the middle, like a kind of Goldilocks star.

And yet a star's power is not just about temperature. Let's again consider the comparison of our sun with the red supergiant Betelgeuse. Astronomers estimate that the temperature of Betelgeuse is around 3,590°K, around 60 percent as hot as our yellow Sun. But due to its gargantuan size, with a radius about a thousand times the Sun's, Betelgeuse is around 100,000 times more luminous. These physical traits mean that Betelgeuse emits a radiant power around 10,000 times as great as the Sun's. So, it seems when you consider not just the star color and temperature of the respective suns, but also their respective energy output, the origin of Superman's powers becomes a little more elusive.

SUNLIGHT SNACKS FOR SUPERMAN

To ponder how sunlight might make Superman snacks, consider coal. Since swimming pools were our entrée to light, there's no good reason coal can't do the same for one of the most vital life forces on Earth: photosynthesis.

Coal is an incredible commodity. People used to call coal "black diamonds" because of its fiery value to humans. In the 1983 *Superman III* movie, the Man of Steel uses his super strength to squeeze coal into diamonds. Now, every basket of coal has brought firepower and civilization. Coal is a mobile climate, as it can be carried to other places, and can make Canada as warm as the Caribbean. Coal's long journey began 358.9 million years ago in the days of the Carboniferous.

The Carboniferous was a period in Earth's history famed for the fact that vast swathes of swamps and forests covered much of the land. During the Carboniferous, the Earth had its highest levels of oxygen in the air, 35 percent compared to 21 percent today. Amphibians dominated the land, and here began the evolution of the first reptiles. But it's the huge plants of the period that made their mark. They grew and died at such a great rate that they eventually became coal. The name Carboniferous means "coal-bearing" and comes from Latin (*carbo* is "coal," and *ferō* is "I carry"). The period was first known as the Coal Measures because of its huge amounts of coal-carrying rocks. Though the Carboniferous started off steamy warm with lots of lush coal forests, it ended up in an ice age that lasted millions of years. By then, the creation of coal had already begun.

The crucial thing about the value of coal is photosynthesis. Billions of years ago, and way before the clammy Carboniferous, plants on Earth learned how to absorb the Sun's energy. Perhaps pause a while to consider the majesty of this ability. Similar to Superman, plants actually *eat* sunshine. They use solar energy to take carbon from the carbon dioxide in the air, and use it to create their living tissue. It is this carbon that burns in a fire, releasing the energy that originally came from sunshine. That's photosynthesis. This sunshine eating does well when it's warmer, and the Carboniferous was sweltering.

The great plants of the Carboniferous eventually died down into the Earth and their energy became frozen in time. Coal preserved sunlight. When we burn coal, the flame of coal fire is the Sun, set free from a fossilized plant. The Carboniferous Earth revolved, and the plants and trees revolved, as the Sun's fiery flame reached the ancient Earth. By photosynthesis the trees converted sunlight into carbon. The tree's roots spread out into the earth to drink its water. Each year the tree grew a new outer-layer, a new ring. Each ring of the many rings of the fossilized tree is one year's worth of frozen Sun-energy, so coal fire is the many years of Sun fire now set free from the tree. Coal is the frozen sunshine of buried forests.

So perhaps this is Superman's secret. Green plants capture only a small percentage of the Sun's energy that reaches them. They then collect that light energy and convert it into what we might call "abilities" inside the plants (photosynthesis kind of means "collecting light," after all). Maybe Superman was able to convert a far larger percentage of the Sun's energy, collecting the light and converting it into his powers and abilities.

BY LIGHT ALONE

Could humans ever use sunshine in a similar way? And what would such a society look like? It's a question that was confronted in the novel *By Light Alone*, written in 2012 by British science fiction novelist Adam Roberts. It's decades in the future, and the world has been utterly transformed by the technology of photosynthetic hair. The advent of the new tech not only means that human hair is able to photosynthesize, it also means that the planet's poor can live, but not thrive, by sunlight alone. Meanwhile, the new tech has swung the pendulum of fashionable elite status display back toward visible public gluttony. Supermodels are now hugely fat and the rich very bald. The world's elite flaunt their wealth and power by cutting their hair, and their unnecessary consumption of food, spending their days eating elaborately ridiculous meals and having back-slapping conversations about how poor people (the so-called "longhairs") are simply lazy and envious.

The photosynthetic hair means a world in which poor people no longer need to work to eat. Global overpopulation mushrooms. An unemployment crisis threatens the lives of the rich and powerful, whose empty lives become more self-absorbed than ever before. They treat thin, longhaired people as social lepers and despise those who work for a living, avoiding the news, which they feel is beneath them. Finally, the novel portrays a nightmarish existence for the world's poor. Their lives, though free from having to find food, has instead plunged them even deeper into poverty and desperation, raising the haunting specter of violence and revolution. Maybe it's not so much fun after all, making like sunshine Superman.

THE MYTH OF THE MASK: WHO IS THE MOST INFLUENTIAL SUPERHERO OF ALL TIME?

CREEDY: "Die! Die! Why won't you die? Why won't you die?"

V: "Beneath this mask there is more than flesh. Beneath this mask there is an idea, Mr. Creedy, and ideas are bulletproof."

—The Wachowskis, *V for Vendetta* screenplay (2005)

"Since mankind's dawn, a handful of oppressors have accepted the responsibility over our lives that we should have accepted for ourselves. By doing so, they took our power. By doing nothing, we gave it away. We've seen where their way leads, through camps and wars, toward the slaughterhouse."

—Alan Moore, *V for Vendetta* graphic novel (1982–1986)

"People should not be afraid of their governments. Governments should be afraid of their people."

—Alan Moore, *V for Vendetta* graphic novel (1982–1986)

"We are told to remember the idea, not the man, because a man can fail. He can be caught, he can be killed and forgotten, but 400 years later, an idea can still change the world. I've witnessed firsthand the power of ideas, I've seen people kill in the name of them, and die defending them . . . but you cannot kiss an idea, cannot touch it, or hold it . . . ideas do not bleed, they do not feel pain, they do not love."

—The Wachowskis, *V for Vendetta* screenplay (2005)

"A revolution without dancing is a revolution not worth having."

—The Wachowskis, *V for Vendetta* screenplay (2005)

"Everybody is special. Everybody. Everybody is a hero, a lover, a fool, a villain, everybody. Everybody has their story to tell."

—Alan Moore, *V for Vendetta* graphic novel (1982–1986)

ANDREA (GALILEO'S STUDENT): "Unhappy is the land that breeds no hero."

GALILEO: "No, Andrea: Unhappy is the land that needs a hero."

—Bertolt Brecht, *Life of Galileo* (1939)

Contrary to common opinion, not all superheroes are American. Shakespeare's sceptered isle has created many memorable heroes over the years: King Arthur, the most mythical of all monarchs; Robin Hood, the most legendary of all outlaws; James Bond, the archetypal superspy; Sherlock Holmes, the definitive consulting detective; Harry Potter, the world's most powerful young wizard; and Doctor Who, hero of the world's longest-running science fiction series. But which superhero, regardless of nationality, has been the most influential in crossing the bridge from fiction and into fact?

A cast-iron case could be made for Superman, of course. One of the most enduring icons in the American cultural landscape, Superman first made an appearance in *Action Comics* in 1938. And yet this powerful fictional figure of the twentieth century has often adapted to changing times. When World War II broke out, his slogan was changed from fighting for "truth and justice" to fighting for "truth, justice, and the American way." This new mantra endured into the 1950s, when Superman became a symbol of a robust American chauvinism, which could do little wrong.

More recently, though, he seems to have adopted the darker moral relativism of the twenty-first century, this brave new world, uncertain and fearful of the fall, whether economic, political, or environmental. In the 2013 movie, *Man of Steel*, after saving a busload of schoolmates from drowning, a distressed teenage Clark Kent confides in his stepfather, who is worried Clark has given away his alien nature. "What was I supposed to do? Just let them die?" Clark says. His stepfather's reply? "Maybe."

It must come as a shock to those who put Superman's popularity down to his embodiment of goodness, of what is right and wrong; the fact that he is a person who is seen to fulfill the same societal function as the myths of ancient Greece or Rome. Some may need myths to teach them virtues. And those virtues may need to be embodied by a person. But if that person truly captures the notion of the Platonic ideal of the good, fighting the fight for high ideals and teaching moral lessons, where on earth is the world going if even Superman sacrifices human life to save his own skin?

For some, Superman is not merely a superhero flying high above us. As an embodiment of our aspirations, our hopes and fears, Superman is us, they claim. And yet consider the 2006 movie *Superman Returns*, starring Brandon Routh, in which an evangelical Christian sits in the White House and there is plenty of chatter about the war on terror being a conflict with Islam. Superman comes closest here to his core depiction, now transformed into a Christlike figure, who descends to Earth from the heavens, and whose true father offers him counsel as he walks among mere humans.

DO THE GODS WEAR CAPES OR MASKS?

And yet social science tells us real change comes from below, not above. Perhaps that's why this uncertain century has embraced a hero who casts off the god-like cape in favor of a mutinous mask. V, created by Alan Moore in his dystopian series *V for Vendetta* between 1982 and 1986, has become one of the world's most recognizable superheroes. V, the masked anarchist, is a mash-up of 1980s British politics and folklore. He has become as iconic as Superman or Captain America, but for very different reasons.

Political protesters worldwide have been wearing V's mutinous mask. The mask is drawn from the movie *V for Vendetta*, and from Moore's graphic novel upon which the film is based. But the legend is older still. The emblematic mask is that of Guy Fawkes, who became synonymous with the Gunpowder Plot to blow up the British Parliament on November 5, 1605, those medieval days of Shakespeare and Galileo. Fawkes's failure has been commemorated in Britain ever since. Fawkes's effigy is customarily burned on a bonfire, usually accompanied by a fireworks display. But the movie told an alternate tale. In *V for Vendetta*, the Guy Fawkes mask is worn by an enigmatic lone anarchist to help stay anonymous.

But in the fiction of Moore's comics and the film, V uses Fawkes as a role model in his mission to end the rule of a fictional British fascist party. Early in the book V blows up the Houses of Parliament, something the actual Fawkes failed to do in 1605.

All across the globe, from New York to London, from Sydney to Rio, waves of protests against politicians, banks, and financial institutions have seen the repeated motif of the strangely stylized Guy Fawkes mask, complete with moustache and goatee. At their peak, over a hundred thousand masks a year were sold. One company even sought to diversify the Guy Fawkes mask image. They designed the traditional Fawkes masks in a variety of skin tones, along with an alternate series of masks with male and female models, with Palestinian kaffiyehs, trees to represent Gezi Park, ancient Egyptian-style ornamentation, and even clown noses, which refer to a long-standing symbol of protest in Brazil. Naturally, some commentators viewed the adoption of this superhero in a cynical way. *The New York Times* noted the irony of an anti-establishment symbol making a royal mint of money and, thanks to its popularity, helping fill the coffers of Hollywood corporation Warner Bros—one of America's one hundred biggest companies with profits each year in billions of dollars.

But the movement goes so much deeper than the mere mask. The drive behind the mask's adoption as a political symbol suggests a seismic shift in the way younger generations view protest and political dissent. The infamous hacker-activist group Anonymous first adopted the mask in 2008 during a protest against Scientology. And Wikileaks founder Julian Assange wore one of the masks during a speech at the 2011 Occupy London Stock Exchange protest. Rumor is that Assange was later forced to remove the mask at the insistence of the police. The subsequent global spread of the mask, often worn wherever there is an emergent protest movement, has been compared to the revolutionary iconography of the famous Alberto Korda photograph of Che Guevara. The superhero V has become the en-vogue revolutionary symbol for the younger generation faced with the recent rise of harsher and more intrusive social media laws. As being recognized by the authorities can lead to a prison sentence, it seems dissenting voices have turned to this mask of anonymity as a symbol of protest.

The man behind the mask is British graphic novel artist David Lloyd. It was he who created the original mask-image for Alan Moore's comic strip, and Lloyd is fascinated by the way his co-creation has become a fashionable symbol for young people across the world. His curiosity led him to visit the Occupy Wall Street protest in Zuccotti Park, New York, to take a look firsthand at some of the people wearing his mask. "The Guy Fawkes mask has now become a common brand and a convenient placard to use in protest against tyranny—and I'm happy with people using it, it seems quite unique, an icon of popular culture being used this way," Lloyd has stated. Lloyd makes the point that the original book is about a lone anarchist bringing down the state. But the film of *V for Vendetta* features a huge crowd of rebellious Londoners, all wearing Guy Fawkes masks, unarmed and marching on parliament. It's this image of collective identification and simultaneous anonymity that is appealing to Anonymous and other groups. In Lloyd's words, "My feeling is the Anonymous group needed an all-purpose image to hide their identity and also symbolize that they stand for individualism—*V for Vendetta* is a story about one person against the system, but the film includes a scene of a huge crowd—making a statement against a faceless corporation."

In the usual way that superhero fiction works, Lloyd says that when he and Alan Moore created the V character, they had the essential notion of an urban guerrilla fighting a fascist dictatorship, but wanted to infuse the story with a strong element of drama and theatricality. Like Deadpool and many other superheroes, V gets his otherness from being a victim of the excesses of scientific testing. Lloyd says, "We knew that V was going to be an escapee from a concentration camp where he had been subjected to medical experiments, but then I had the idea that in his craziness he would decide to adopt the persona and mission of Guy Fawkes—our great historical revolutionary."

The V masks were initially made by Warner Bros to promote the movie, and were given away at screenings. Few people could have predicted the sales of the masks would mushroom into a global phenomenon. Lloyd says he has heard rumors of US police searching for the masks in people's houses to be used as evidence of association with Anonymous and political activism against corporate greed.

All this makes V quite an exceptional hero compared to most others. It's hard to imagine the authorities ransacking your crib in search for evidence that you're hiding a Superman cape, or a Captain America shield. But given that the movement is all about bypassing governments and starting from the bottom, Alan Moore's declaration, "people should not be afraid of their governments. Governments should be afraid of their people," has to some extent proved to be very prescient.

PART II
TIME

DAILY DIARY: THE NEED FOR SPEED—HOW THE FLASH DEALS WITH DYNAMICS

"It is not by muscle, speed, or physical dexterity that great things are achieved, but by reflection, force of character, and judgment."
—Marcus Tullius Cicero

"Speed is often confused with insight. When I start running earlier than the others, I appear faster."
—Johan Cruyff

"The dragonfly is an exceptionally beautiful insect and a fierce carnivore. It has four wings that beat independently. This gives it an ability to maneuver in the air with superb dexterity. A dragonfly can put on a burst of speed, stop on a dime, hover, fly backward, and switch direction in a flash."
—Richard Preston, *The New Yorker* (2012)

A picture of tranquility. Or so it seems. It's the opening scene of the 2013 movie, *Gravity*. A section of planet Earth appears as the camera tracks slowly and serenely about our world. A space shuttle becomes visible, as it approaches the limb of the Earth. Soon, an astronaut comes into view from behind the shuttle, and he too seems to coast coolly above the blue marble of the globe below.

And yet the scene is typical of one of the most deceiving sights in nature. The astronaut, along with his space station and the Hubble Space Telescope they soon repair, are orbiting the Earth every ninety minutes.

Yep, every one and a half hours, that super astronaut encircles our entire world. And that means he is traveling at a speed of 17,500 miles per hour. Way back in 1941, Superman was said to be *"faster than a speeding bullet."* Well, modern astronauts, moving in orbit at a speed of about 5 miles per second, are doing "a Superman" with ease. Indeed, at that speed, the astronaut would cross the length of a football field before the bullet had gone 30 feet.

You don't even need superheroes to move faster than the wind. So, what's the catch? Think about the very plot of the *Gravity* movie. Local space simply isn't as clean as it used to be. During that spacewalk to service Hubble, Mission Control in Houston warn the astronauts that a missile strike on a defunct satellite has caused a chain reaction, forming a cloud of debris in space. Their mission must be aborted immediately, as the debris is hurtling towards the Shuttle, faster than speeding bullets!

We get the briefest glimpse of how dangerous this is. [SPOILER!] While trying to return to the shuttle and open the airlock, the crew's flight engineer was killed when a chunk of debris the size of a softball smashed its way through his helmet, tore through his head, and exposed him to the vicious vacuum of space.

The same sort of danger would face The Flash. Air is not empty. Elements such as nitrogen (78 percent) and oxygen (21 percent), and myriad microscopic dust particles, make up the atmosphere around us. When you move past these elements in the air, you rub up against them. And that creates a lot of friction, which ends up as heat. You can get an idea of this just by rubbing your hands together. The friction between your palms warms them up, and, the faster you rub, the more heat is generated. At some point in the ancient past, it's the way our ancestors made fire from rubbing sticks!

THE NITTY-GRITTY OF SPEED

Say you're as slick as Quicksilver. You're moving at 20,000 mph. At that speed, the heat from friction would be enough to burn your face off. Even if you were able to withstand the heat, there's still those tiny and gritty air particles to deal with. That dirt and sand in the atmosphere would make millions of little lacerations to your form, as you race by. Sure, you've seen the state of a windscreen that's been on the road for some time. Imagine what those bugs and beetles would do to your face and body, let alone the bird hits! Little wonder Flash is so fussy about his mask and suits, but you have to worry about Quicksilver's costume, as in the movie, *Avengers: Age of Ultron*, it seems little more than a long-sleeve tee-shirt.

And what about the built environment between you and your mission? Scholars say it takes humans about 0.2 of a second for us to react to something we see. So we can work out how swiftly The Flash and Quicksilver would have to be reacting to dodge and dart between the buildings in their path. Assume that Quicksilver is again moving at a speedy 20,000 mph. That means in one minute he's moving over 333 miles, and in one second he's going 5.5 miles (that's only a little faster than the space station we mentioned earlier). So, in 0.2 of a second, the average human reaction speed, he'll have travelled 1.1 miles. Before he's even reacted to something in his path, he'll have shot past it by over a mile! He'll either get totally mashed by crashing into the nearest wall at super speed or, if he also has indestructability as a power, transform his careering frame into a weaponized missile, annihilating everything in his path!

In short, a high-speed human, armed with otherwise normal human abilities traveling long distance at 20,000 mph would end up a charred mess, caked in beetles and bugs, and careering forward with little hope of avoiding the next crash into catastrophe.

Okay, what about dragonfly tactics? How about short bursts of speed, to a location you can see, and with no buildings in between? Imagine a speeding bullet, going no faster than a space station, is about to plow into yet another damsel in distress. You, our speedy hero, swoop to the rescue at super speed. You lunge for the damsel, snatch her safely from harm's way, and swish her swiftly to sanctuary. Sounds good, if a little clichéd. And yet the truth is the damsel would probably suffer more damage from her beau than the bullet he helped her dodge.

This "dragonfly" dynamic all depends on inertia. Now, inertia is the resistance to a change in the state of motion of a body. Newton nailed inertia in his First Law of Motion, which we can reword in an everyday way as, "A body at rest will stay at rest, forever, as long as nothing pushes or pulls on it. A body in motion will stay in motion, traveling in a straight line, forever, until something pushes or pulls on it."

That "forever" part is a bit of a hard sell. But think again about the movie *Gravity*. Sandra Bullock plays an astronaut who becomes adrift in space. Once her body is in motion, her inertia *will* keep her going forever into the depths of space. But her character uses the spray of the fire extinguisher to act as a thrust to pull her back on track.

Let's go back to that damsel bullet dodge. So, as the bullet approaches, the damsel will stay put unless something changes her condition. That means she is at rest, traveling at zero miles per hour. As the hero swoops in to the rescue, traveling at Quicksilver pace, the damsel's speed surges rapidly to 20,000 mph. The outcome? The damsel's brain would smash into the side of her skull. Likewise, if our hero stopped suddenly from Flash speed with the damsel in his arms, her speed would decrease rapidly back to rest. And that would also mean her brain would smash into the side of her skull, turning her gray matter into red Jello. Come to think of it, the rest of her internal organs might suffer from a similar complaint. In short, and in terms of physics, what our speed hero does to the damsel is effectively the same as a space shuttle careening into her. She no doubt dies at the moment of impact.

So, here's a lesson if you want to save lots of time by being able to move fast and with dragonfly dynamics. Those mid-air maneuvers of superb dexterity, that ability to stop on a dime, and those switches of direction all in a flash? They take the refinement of millions of years of evolution. The same is true of the super-powered mantis shrimp, which travels so fast through water that the sea ahead of it boils as it passes. Or the Portuguese man-of-war, whose poison stings are released in 700 billionths of a second—the fastest known animal mechanism on Earth. It takes more than a quick fix superpower to be a master of speed.

HOW DOES USAIN BOLT COMPARE WITH CAPTAIN AMERICA?

Exterior. Washington, DC. Sam Wilson is jogging when Steve Rogers quickly catches up to him and runs past him.

STEVE: "On your left."

[AS SAM CONTINUES TO JOG STEVE COMES AROUND AGAIN QUICKLY AFTER DOING ANOTHER LAP]

STEVE: "On your left."

SAM: "Uh-huh, on my left. Got it."

[AS SAM IS STILL JOGGING STEVE COMES AROUND AGAIN FROM BEHIND HIM FROM ANOTHER LAP]

SAM: "Don't say it! Don't you say it!"

STEVE: "On your left!"

SAM: "Come on!"

[SAM GETS ANGRY AND TRIES TO CATCH UP TO HIM BUT ONLY AFTER A FEW SECONDS HE'S UNABLE TO CARRY ON AND STOPS TO REST. AS SAM IS RESTING CATCHING HIS BREATH SITTING BY A TREE STEVE WALKS OVER TO HIM]

STEVE: "Need a medic?"

[SAM LAUGHS]

SAM: "I need a new set of lungs. Dude, you just ran like thirteen miles in thirty minutes."

STEVE: "I guess I got a late start."

—Christopher Markus and Stephen McFeely, *Captain America: Winter Soldier* screenplay (2014)

"I am a living legend. Somebody said if I win these three gold medals I would be immortal and I kind of liked it. So I'm going to run with that one."

—Usain Bolt at the Rio 2016 Olympics, quoted in *The Daily Telegraph* (2016)

"*I am Bolt* stripped away the superman packaging, and showed an athlete forcing himself to go on winning while also being desperate to stop."
—Paul Hayward, chief sports writer at the *Daily Telegraph*, on Twitter (2017)

"The serum amplifies everything that is inside. So, good becomes great. Bad becomes worse. This is why you were chosen. Because a strong man, who has known power all his life, will lose respect for that power. But a weak man knows the value of strength, and knows compassion . . . Ladies and gentlemen, today we take not another step toward annihilation, but the first step on the path to peace. We begin with a series of micro-injections into the subject's major muscle groups. The serum infusion will cause immediate cellular change. And then to stimulate growth, the subject will be saturated with Vita-Rays."
—Dr. Abraham Erskine in *Captain America: The First Avenger,* Christopher Markus and Stephen McFeely screenplay (2011)

The most famous race in the history of the Olympics is also its simplest. Eight of the planet's most primed sprinters sit at the blocks of eight straight tracks. A starting pistol fires. An explosion of muscle, sinew, and tendon and, a mere ten seconds later, an Olympic champion. The short and sharp simplicity of the race: just pure sprinting power. No jetpacks, no gadgets or gimmicks, no superhero suit—just raw and natural speed. And yet, in those fleeting moments of the race, the 100-meter athletes perform physical feats so complex that scholars are still trying to fathom them.

You'd think it would be easy enough. Famous British scientist Sir Isaac Newton started piecing together the mechanics of motion back in the late 1600s. But, when those laws of motion are applied to complex biological systems, such as the human body, the applied physiology and biomechanics of the case makes things a lot more complicated.

We humans have just gotten quicker and quicker over the last hundred years or so. The men's 100-meter sprint has been on all Olympic Games programs since the first modern Olympics in 1896. The Olympic sprint champion is hailed the fastest man on Earth, and the whole world seems to stop to watch it. US athletes have dominated the race, winning more times than any other country, with seventeen out of the twenty-seven times that it has been run. And those finishing times have gone from the 12-second winning time of American Tom Burke in 1896, to the US's Justin Gatlin's 9.85 seconds in 2004. Then Usain Bolt happened, and the home of the supreme sprinter moved from the US to Jamaica.

The human brain is the unsung hero behind all great athletes. And in the 100-meter race, the brain switches to autopilot. When sprinters are in place at their starting blocks, they must mentally screen out all other thoughts. Their main task is to be steady and focus. But when the pistol triggers, the brain sends instant word to the muscles, and all becomes automatic. The sprinter's body stays low, the muscles contract to create the force needed to kick back against the starting blocks. Time to accelerate.

Each footfall on the track brakes the motion a little before pushing forward once more. For the very top athletes, footfall contact is quicker than an eye blink. When the race is halfway done, the elite sprinters reach maximum velocity. In the case of Usain Bolt, this speed is almost 30 mph. Now the footfall force that hits the track is huge—more than triple the sprinter's body weight. And only seconds later, when the race comes to a close, energy stores are run down, and pace begins to fade. When it looks like the Olympic champion is pulling away from the pack, remember this: they're simply slowing down the least.

Research has begun to show that what makes these sprinters fast is how hard their footfall impacts on the track, compared to their body weight. It may look like Bolt is gliding along the track, but in reality, it is his sheer power that counts. When the race's replay cuts to slow motion, you begin to see the truth behind the grace. The slo-mo footage shows the rippling in the flesh of the faces of the athletes, a telltale sign of the force impacted from foot to floor. This footfall force is what makes Bolt so supreme. Bolt's footfall maxes out at an incredible five times his weight. Lesser athletes peak at about 3.5 times. And so, the science suggests top sprinters simply deliver the greatest footfall force to the track, and that's what makes them elite.

THE CAPTAIN VERSUS THE BOLT

We could compare the Olympian to the Captain in terms of sheer speed. (Forget the likes of the Flash and Quicksilver, and their silly pseudoscience speeds). Bolt's peak speed during his 2009 world record run in Berlin was 27.8 mph, not far off a city speed limit. And, according to Sam Wilson, Steve Rogers's speed was, "like thirteen miles in thirty minutes." At that pace, the Captain would clearly clock twenty-six miles in an hour, which would cut the world record marathon time by more than a half. But we're clearly not comparing like with like here, as the Bolt story is all about pace, whereas the Captain's tale is all to do with stamina. So, far better to think about the race in terms of the comparative force of footfall.

Let's run the race. Trial one: Two of the planet's most primed sprinters sit at their blocks. In lane one, Usain Bolt a.k.a. "The Bolt," the first athlete to hold both the 100-meter and 200-meter world records since fully automatic timing became compulsory, and widely considered to be the greatest sprinter in history. In lane eight, Steve Rogers a.k.a. Captain America, patriotic super-soldier, ranked sixth on IGN's "Top 100 Comic Book Heroes of All Time" in 2011, second in their list of "The Top 50 Avengers" in 2012, and second in their "Top 25 Best Marvel Superheroes" list in 2014.

At the starting blocks, the brains of our heroes try to block out all other thoughts. Bolt tries not to wonder whether he should have had more Chicken McNuggets—he once powered to three gold medals fueled by the food, estimating he had consumed one thousand McNuggets during ten days at the 2008 Beijing Olympics. The Captain also tries to focus, and put out of his mind whether he should arrest the race official for possession of an offensive weapon. The starting pistol fires—the same explosion of muscle and might transformed into sheer sprinting power. Once more, no jetpacks, no gadgets or gimmicks—just the Captain's stealth suit, the outfit with a more tasteful and subtle stars and stripes motif.

Now the comparative force of footfall comes into play. That telltale sign of the force impacted from foot to floor maybe supreme with Bolt, but it must surely be even better with Rogers. That series of micro-injections into his major muscle groups; the serum infusion that caused instant cellular change; and the stimulated growth from the saturation with Vita-Rays. We can assume that the Captain's footfall force will be greater even than the fives times body weight that Bolt belts out. And given that Rogers can keep his pace over those legendary thirteen miles, when the short race draws to a close, and energy stores run down, it's the Captain's pace that is least likely to fade.

What other clues could there be to the Captain's supremacy? The key to Steve's speed would be to enhance his footfall force without gaining body weight. Top sprinters hit a peak of four or five times body weight in an incredibly short time after leaving the blocks, and their footfall is less than a tenth of a second on the floor, three times faster than the blink of an eye. And to generate such forces, the Captain would need to turn his footfall steps into precision punches.

Another factor in Steve's performance would be fatigue. Human sprinters not only have an ongoing fight with mental focus, they also have to fight against fatigue, which starts almost instantly. If those Vita-Rays have any other effect on sprinting, it would be an enhancement in the mechanics and chemistry of tiring. Experiments done on stationary bikes, where a large resistance is added suddenly to a rider peddling at high cadence, shows that the power exerted falls on the second stroke, and continues to drop. It's the same on the track. A great advantage for Bolt is his stride length. It means he runs forty-one steps in a race, compared with forty-five for his opponents, so his muscles have four fewer impacts to get tired. Steve would at least have to match Usain for stride rate.

Finally, the most common question sport scientists are asked is how fast can humans run one hundred meters? If Steve could maintain Usain's footfall force, while also cutting contact time with the ground to just seventy milliseconds (down from about eighty), he would hit a top speed of 28.53 mph—and a new world record of 9.27 seconds. Some clues as to how the Captain might do that come from champion Finnish cross-country skier Eero Mäntyranta. Mäntyranta had a genetic defect which meant his red blood cells could carry more oxygen. He was, essentially, an X-Man and a natural doper. Not Vita-Rays, but plain old O_2.

THE CAPTAIN, THE BOLT, AND THE CHEETAH

But before we all get carried away with superhuman supremacy, consider the humble cheetah. We are all fairly familiar with this big cat, but let's remind ourselves of some feline facts. The cheetah occurs mainly in eastern and southern Africa, and some areas of Iran. The cheetah is known for its slender frame, deep chest, yellowish-tan coat covered with nearly two thousand solid black spots, small round head, and long thin legs. Its slight and slender form is in sharp contrast with the more robust build of other big cats. But boy can it shift. A cheetah's top speed is almost 70 mph. Imagine that: you could be coasting along the highway and this most elegant of creatures would still leave you behind in its dust.

Let's run another virtual race. Trial number two: This time we'll put the cheetah in lane four, between the Captain and the Bolt—time for some real fierce competition. The cheetah's dominant asset is its ability to accelerate like a rocket. So, as the race begins the cat takes off from the blocks and is swiftly ahead of the humans, not only finishing in 5.9 seconds, but also beating Bolt's world record time by more than 3.5 seconds. Not even the Captain can match her pace.

Trial three: We give our human and superhuman runners a head start. If we hold back the cheetah in its starting block, and let the humans race for the first 40 meters, that would leave the last 60 meters to race. As the humans hit the 40-meter mark, we release the cheetah, whose incredible acceleration means she catches up, and all three of our sprinters cross the finishing line at the same time.

The secret to the cheetah's speed? A sleek body packed with more "fast twitch" muscles, muscle fibers built for quick bursts of power. Mere humans have 'slow twitch' muscle fibers, which, as with the case of the good Captain, enable stamina and endurance, but are not so good with sprinting. Secondly, there's the cheetah's spine. Her back is bent like an archer's bow, which springs her headlong the optimal distance with each stride. Once set off, she is halfway to the human's position in just a single second. And as the humans reach the race's end, she's literally becoming airborne with each seven meters stride, gaining speed as she finishes. Little wonder the cheetah wins in about half the time.

WHERE DOES AQUAMAN SIT ON THE TREE OF LIFE?

"[Evolution on Earth was] . . . represented in the old charts and texts as [a ladder of] an 'age of invertebrates,' followed by an age of fishes, age of reptiles, age of mammals, and age of man (to add the old gender bias to all the other prejudices implied by this sequence) . . . We will not smash Freud's pedestal and complete Darwin's revolution until we find, grasp, and accept another way of drawing life's history. JBS Haldane proclaimed nature 'queerer than we can suppose', but these limits may only be socially imposed conceptual locks rather than inherent restrictions of our neurology. New icons might break the locks. Trees—or rather copiously and luxuriously branching bushes—rather than ladders and sequences hold the key to this conceptual transition."

—Stephen Jay Gould, "The Evolution of Life on the Earth," *Scientific American Life in the Universe Special Issue* (1994)

"I also have a hairline fracture in my thumb. Mankind's least important finger; am I right?!"

—Detective Jake Peralta, *Brooklyn Nine-Nine,* episode AC/DC, (2015)

Where does Aquaman sit on the tree of life? He may not be everyone's favorite fishy super-fella. But Aquaman has staying power. Since his creation way back in 1941, he has been among DC's most durable superhero icons. During the Golden Age of comic books (from the late 1930s to circa 1950), Aquaman held out against Batman, Superman, and Wonder Woman. And in late 2017, Aquaman hit our movie screens as one of the members of the Justice League, who join forces to face the imminent threat of Steppenwolf, and his army of Parademons, who are heading to Earth, naturally. In 2018, Aquaman even has his own superhero film, based on this "king of the undersea nation of Atlantis," whose superhuman aquatic abilities and herculean physical powers stem from his Atlantean physiology.

Aquaman's powers were tied to the aquatic realm. In the early days, when Superman's powers were positively endless, and Batman's technological brilliance was peerless, Aquaman had to be more than just a superhero; he had to be a little more human. But on what branch of the human family tree does this compelling and complex hero rest? Let's consider our evolution as humans, and see where our aquatic cousin might fit in.

READY PLAYER: HUMAN EVOLUTION

What if we made a strategy video game of human history, one that enabled us to replay through our evolution?

To set our world-game in motion, we have to start human civilization all over again from scratch. How far back would the game have to begin to re-spawn civilization? The answer is likely to be way back in time, before ancient Romans or Egyptians, and deep in what scholars call pre-history. Indeed, the probable answer is our world-game begins with the Stone Age.

The Stone Age is the age of early humans. Not just today's modern humans, which are known by scholars as homo sapiens ("wise man"), but the human ancestors that came before us too. The Stone Age is the entire period from when our human ancestors developed, and evolved into intelligent creatures that learned to make and use stone tools to survive. It would be some strategy game!

Our human evolution video game would need to embrace the human timeline, as best we know it so far, drawn up from what we've found out about the evolution of humans from fossils and the like:

10 million years ago: Early apes roam the Earth
6 million years ago: Human-like apes develop
4 million years ago: Earliest bipedal
2.8 million years ago: Earliest stone tools, homo habilis ("able man") appears
1.8 million years ago: Earliest exit from Africa, homo erectus ("upright man") appears

1.4 million years ago: Earliest use of fire
0.25 million years ago: homo neanderthalensis ("Neanderthal man") appears
0.195 million years ago: homo sapiens ("wise man") appears

In our strategy video game of human evolution, our players would need to eat to survive. Now, modern shopping is a bit of a breeze, isn't it? All you need do is drive down to the supermarket, and boom, shelves full of fruit, vegetables, meat, and fish. But in our strategy video game, when we start civilization from scratch, our players have to hunt and gather all their own food. They have to seek for themselves the riches the Earth has to provide. Prehistoric people had to kill to eat meat. Stone Age tools were a vital development in our evolution. And hunting became truly significant, as is witnessed by the fact that ancient cave artists made it the main theme for many of their rock paintings.

The video game would recognize that Stone Age humans hunted mammoths, reindeer, and buffaloes. Through custom and practice, prehistoric people made a very important discovery: cooperation and people power made for an easier hunt. Scholars and commentators too often stress the selfish, individual aspects of evolution. But the truth is that homo sapiens found there is far more power in cooperation and numbers—the more people involved in the hunt, the better the hunt went. And this was especially true if they planned the hunt. As they had little access to sawed-off shot guns, Uzis, and photon torpedoes, planning, thinking, and sheer stealth was far more important than sheer force. That would have to be programmed into our game too. What weapons they did have would have been the stone weapons of the earth, which were also crucial (one can easily picture our strategy game's weapons menu now, with the options for sawed-off shotguns, Uzis, and photon torpedoes grayed out, but the choice of "primitive hand axe" freely available).

READY PLAYER: THE AQUATIC APE THEORY

But let's rewind a moment and consider hairy apes. How much hair cover would we give our characters in this strategy video game of human evolution? Many mammals have a layer of thick body hair, which they use as insulation to keep them warm. However, modern humans don't have this *visible* layer of hair. (Meanwhile, aquatic mammals evolved thick layers of fat to help normalize their body temperature). An adult human has around five million hair follicles, distributed quite evenly across most of the body. For scholars, it's little surprise to learn that the number and density of our hairy frame is similar to our nearest primate, the chimpanzee. (One should desist from mentioning this too often to one's partner in romantic situations, in my experience, as modern human females, in particular, resist the suggestion they are as hair-covered as other apes.) The major point of divergence is the thickness, length, and darkness of the hairs.

We're the abnormal apes. But how aquatic are we? The evolution of this divergence in ape hair cover has many scientists wondering. Of all the existing primates, we're the only species to have lost that thick coat of hair. Most of our hair follicles produce small and unpigmented vellus hairs, rather than thicker, pigmented terminal hairs, such as the hair that typically grows on the scalp. Despite being as hairy as chimps, humans appear quite naked.

Conventionally, it's believed that bipedal walking was a necessary step to our transition from a dense to sparse hair covering. And it was normally believed this transition took place more than one million years ago on the African savannah, with homo erectus apes. Scholars figured that for humans, developing in the hot open African savannah, physical activity was the least heat exhausting during the cooler parts of the day. As our body hair evolved and sweating improved, humans were able to extend the time they were active during the day. And that would give our game characters an advantage for hunting food, getting water, and doing anything else needed to survive. But perhaps our strategy video game of human evolution should also allow for the idea of a more aquatic ape.

Counterpoint: the Aquatic Ape Theory, a.k.a. AAT. The controversial theory that humans evolved from amphibious apes has won over a number of scholars, including British naturalist Sir David Attenborough, since it first emerged fifty years ago. The original theory was developed by Sir Alister Hardy, a British marine biologist, who was an expert on marine ecosystems, spanning organisms from zooplankton to whales. Late in his career, Hardy gave a talk at the British Sub-Aqua Club, and a month later published an associated article in *New Scientist* on the same topic, "Was Man More Aquatic in the Past?" Hardy's paper presented most of the basic ideas, and the method, of the AAT, which is that human apes emerged from water, lost their fur, started to walk upright, and then developed big brains.

Another supporter of the AAT was prominent TV author Elaine Morgan, who also wrote several books on evolutionary anthropology. Morgan, an Oxford graduate in English and a resident of my hometown of Mountain Ash, entered the scene in 1972 with her book, *Descent of Women*. This pioneering work, first published in 1972 and revised in 1985, was the first to argue, intelligently and indubitably, the equal role of women in human evolution.

Morgan followed up *Descent of Women* with *The Aquatic Ape* in 1982. She became drawn to the topic when reading about the savannah hypothesis of human evolution in Desmond Morris's book, *The Naked Ape*. Morgan was irritated by the sexist explanations, which she felt were largely male-centered. For example, if humans lost their hair because they needed to sweat while chasing game on the savannah, that didn't explain why women should also lose their hair, as according to the savannah hypothesis, they would be looking after the kids!

The AAT was given a boost when the savannah hypothesis of human evolution ran into further trouble. Previously, the savannah hypothesis had been hugely cited as the primary developmental vector of the evolution of ape into human. It was even touted as the main motivating force for going from four limbs to two, as the upright gait was alleged to have given its perpetrators the advantage of gazing over the tall savannah grass, watching out for predation from big cats, such as the lion or sabre-tooth.

But fossil find after fossil find raised major doubts. The archaeological evidence in the later years of the twentieth century showed that, millions of years ago, early humans were to be found in regions of the Earth that had been wood or forest, and not savannah. And the idea of the savannah itself was no longer as originally envisaged. New studies suggested that early humans were ill-suited to the savannah: our cooling-down system doesn't do well in exposed climates, such as the savannah; we have far too many sweat glands; we waste both sodium and water; and we simply can't take on enough water at a single time, a criterion critical for a savannah living. Consequently, scholars mostly dumped the savannah hypothesis in the mid-1990s.

A BRIEF WORD ABOUT WATER

Before we continue with strategy video game of human history, consider Planet Earth. Or should we call it Planet Water: the Home of Aquaman?

Water has the greatest power over our world. For many people, water is the most magical force on the planet. Water is the world's lifeblood. It shapes the contours of the land. It renews and recycles our weather systems. And it nourishes, by coursing through nature with the crucial ingredients for life. Water is made up of the explosive ingredients of hydrogen and oxygen. And yet together they make water, a harmless change agent, which shapes mountains, canyons, and coastlines. By the unleashing of hurricanes and floods, water's destructive power is awesome. It has shaped our lives in a way that's seldom said in history books. Water makes Earth alive.

On a planet where water is so prominent, perhaps it's no surprise that scholars believe life was born in water. As the saying goes, "*Water is life's mother and medium; there is no life without water.*" The word "matrix" means a habitat in which life can develop and thrive. Water is a perfect matrix. It is a calm and neutral substance, and it stays as a liquid over a wide range of temperatures. But water is unique in other ways too. It makes 3D bonds with other chemicals, and is an active change agent, which means water makes things happen. Water enables the kind of delicate chemistry that made life possible. And that goes back to the very origin of life itself.

Because water is needed for all life, scholars believe that life on Earth probably began in our oceans, around 3800 million years ago. From there, life very slowly took hold on land, but that didn't happen until about 450 million years ago. It seems life was very happy in water, and didn't want to leave in a rush.

All the wonderful variety of life on Earth depends on access to water. Most plants are 80–90 percent water. The banana tree is 90 percent water. Water is essential for seeds to germinate, and for plants to grow. Water allows plants to absorb nutrients from the soil, and move nutrients between cells. Water also provides the pressure in plant cells, which enables them to stand upright.

READY PLAYER: AQUATIC HUMANS

Here's something new to key into our strategy video game of human evolution: gameplay cuts from savannah scene to aquatic habitat. Evolutionary studies hold that humans, gorillas, and chimps share a common ancestor. But, whereas gorillas and chimps have numerous common traits, human traits strongly suggest a divergent evolutionary journey. The AAT tries to explain how humans became very different from other apes. We lack (obvious) fur, walk upright, have big brains, subcutaneous fat (more on this soon) and have a descended larynx—which is common among aquatic animals. The AAT says that, a few million years ago, we diverged from the other apes on a different evolutionary path because sufficient numbers of our ancestors found themselves having to survive for many millennia in flooded, semi-aquatic habitats.

How did we get naked? Even though humans are primates (the order of creatures that incorporates monkeys and apes), among the many primate species, only humans are nude. And the main types of habitat that make for naked mammals are either underground or aquatic. All the other mammals, with sparse or absent fur, such as dolphins and whales, spend huge swathes of time either in water totally, or basking constantly in mud, like elephants and pigs.

A further factor on the route to the naked ape is the question of subcutaneous fat. The word subcutaneous here essentially means "under the skin." Fur may be fine insulation for terrestrial mammals, but the best insulation for aquatic mammals is subcutaneous fat. Modern humans are the fattest of the primates (another fact best not mentioned in romantic situations), with ten times the fat cells around our frames as should be the case for a primate of our size. How did we get that fat? Think much less about prehistoric burger bars and more about hibernation and water. The main animal types for possessing large reserves of fat are those that hibernate, and those that dwell in water. Like aquatic creatures, humans store fat mostly under the skin. Not so with terrestrial mammals. They store their fat internally. And since it's unlikely humans developed their subcutaneous fat living and surviving on the savannah (fat guys have more trouble hunting), it's more likely we accumulated our fat when we dwelled near water.

And how about "two legs good, four legs bad"? We are the only mammals that have gone permanently bipedal. And it's no easy option. Walking upright as we do is far trickier than walking or running on all fours. Scholars once thought we first got big brains, maybe from using tools and weapons, and *then* started walking upright. But fossil finds soon showed this theory was the wrong way around: we went bipedal before we went big brained. But why go bipedal if it's harder than naturally being on all fours?

Aquatic apes may well provide an answer. Scholars of the AAT say that an aquatic or flooded habitat would have forced prehistoric humans to "go bipedal." Keeping your head above water seems a pretty essential habit for survival, and walking on hind legs is a good way to do it! And we have contemporary examples of the same thing. There are a couple of monkey species that are sometimes seen to go bipedal. They're forced to do so when their swamp-like habitats become seasonally flooded, and walking on two legs is the way to go.

So our strategy video game of human evolution will need an aquatic option. Scholars are skeptical that the AAT explains all we know about the evolution of early humans. And yet it's a very compelling theory in many parts. The AAT supposes that the first humans arose from a seriously flooded Africa, about 5 million years back in history. This revolutionary and abrupt shift in habitat is thought to have forced us to evolve along a new developmental path. Our early journeys since then are also intriguing.

READY PLAYER: BUSY BEACHCOMBING

Take, for example, the early modern human migration out of Africa. It has strong aquatic features. Our strategy video game of human evolution should detail genetic evidence, which suggests the human population around seventy thousand years ago had crashed to around only two thousand individuals. Humans nearly went extinct. We were hanging on by our Stone Age fingernails (surely an exciting prospect for a video game, with any failure in gameplay resulting in a "HUMANS EXTINCT!" message outcome, flashing bright red across your screen). But even though the ice age made hunting hard, a small group of us left Africa and ended up in Australia.

How do scholars know humans travelled to Australia? And how did they get there, when there is so little archaeological evidence of their journey? The answer to the first question is easy enough: Around 100 million years ago, Australia separated from Pangaea (the name given to the one large area of land on ancient Earth, which began to break apart about 200 million years ago to create the present continents, such as Australia). All the plants and animals on the Australian continent are different. There are no primates. And so humans must have come to Australia from somewhere else. Humans had to be an African import.

Scholars estimate that from a population of two thousand to five thousand individuals in Africa, only a small group, possibly as few as 150 people, crossed the Red Sea. Though the Sea never completely closed during that ice age, it was narrow enough to cross and there may have been islands to enable crossing using rafts (cue Red Sea Raft section of the strategy video game, where our players brave the waves to get to the distant shore). The travelers journeyed eastward, out of Africa. Within a few thousand years, the climate became drier, making it harder to turn back.

But here's the thing: their migration route from Africa to Australia was aquatic. The most obvious journey was along the coast of southern Asia, as the coastal route has no big change in climate or environment. Our intrepid travelers, and their descendants, took this beachcombing coastal route, reaching present-day Malaysia within a few thousand years. And, by 45,000 years ago, humans were living in parts of Australia. There's archaeological evidence that Aboriginal Australians have been on the continent for that long. They made campfires, which can still be detected. And they made their mark in cave art.

READY PLAYER: AQUAMAN

So here's a possibility: Aquaman is a turbo-charged aquatic ape. Scholars of the AAT say that certain human features in anatomy and physiology, which are only seen in aquatic creatures and humans, are proof that our ape ancestors went through an aquatic phase in their transition from ape to hominid. A common theory of evolution is the question of convergent evolution: the appearance of apparently similar structures in organisms of different lines of descent. Life in an aquatic environment explains certain human features, which a transition from ape to hominid in a non-aquatic environment cannot. So, our ancestors came down from the trees to live in the food-rich creeks, rivers, and seas. We evolved to become upright as we tried to keep our heads above water, and lost our hair. We developed fat to keep warm in the water.

We could use convergent evolution to conjure up Aquaman also. The AAT says that modern humans went bipedal and got big-brained by adapting to their partly aquatic habitat. It could easily be that the race of Aquaman developed gills. Remember that human embryos go through a stage where they have slits and arches in their necks like the gill slits and arches of fish. These structures do not develop into gills in humans, but the fact that they are so similar to gill structures in fish at this point in development supports the idea that humans may share a common ancestor with fish.

Our strategy video game of human evolution suddenly goes all Gungan. Like those alien creatures from *Star Wars*, a human race of Aquaman could have developed gills and learned to live underwater, adapting to a submarine habitat like those Gungan guys on Naboo. Aquaman is, after all, a marine mammal. And, as he's effectively hairless, we'd seriously need to blubber him up for our strategy game. Typical human body temperatures sit at around 37°C. Even the warmest waters of Earth's oceans are only around 27°C. And in the deep sea, at what we might call Gungan depths, the ambient temperature drops to about 2–4°C. In short, the ocean is cold. And warm-blooded species would need to evolve heavy layers of blubber to cope.

Meanwhile, Aquaman is seriously ripped. In some of the comic book characterizations of him, Aquaman looks to have less than 5 percent body fat. This guy ain't blubbered enough. To give better measure, warm-water bottlenose dolphins have up to 20 percent body fat. And, as anyone in the British Sub-Aqua Club would have freely told Sir Alister Hardy had they possessed the kit, even decked out in a 12 millimeter neoprene wet suit, a few hours wading around in ocean waters will get you arctic. Aquaman, lacking any visible sign of such a suit, is more likely to contract hypothermia than the attention of the Justice League. So our strategy Aquaman character needs to get adiposed.

Then there's the pressure of living submarine. The human mammal is hypotonic compared to seawater. In other words, you find more molecules in seawater than you do in our cells. So, if Aquaman is lugging seawater into his lungs, and other air chambers, he has to keep an equilibrium in his body, for his cells to work as they should. So, his cells start to flush out water to increase their molecular density. As he loses water, his cells shrivel and his internal organs start to fail, beginning with his kidneys, and end up with his circulatory and respiratory systems failing (cue video gameplay resulting message outcome of "AQUAMAN EXTINCT!" flashing bright green across your screen). But let's give our game the benefit of scientific doubt. Let's provide our Aquaman strategy character with evolved ways of coping with the aquatic challenges of temperature and pressure (plus the problems with "the bends"—or decompression sickness—that we talk about in this book's entry on how Iron Man and Superman cope with the physics of flight).

And, finally, the din of it all—the pandemonium of the deep. What kind of aural gameplay experience should we inflict upon our strategy Aquaman character? Much has been written about the psychology of sound in video games, and the way in which sound can play with the gaming mind. The right musical score or sound effect can conjure up the kind of emotional reaction that moviemakers have been coining in on for decades. So sound is a powerful asset in the developer's toolkit, which will enable the kind of immersive and emotionally-charged ambience Aquaman can expect under water. And that ambience is the relentless, unceasing screams of dying marine life.

The main noise layer of our soundscape of the deep would be the shrimps. The shrimp layer is the loudest thing in the ocean. Trillions of shrimp bob about en masse, creating bubbles, which pop on such an industrial scale it not only keeps people awake, but also whites out submarine sonar systems, and deafens sonar operators through their headphones. Submarines sitting at shallower depths than a shrimp layer can't hear anything below it, and vice versa. The noise of these hordes of shrimps reaches 246 decibels, the equivalent of 160 decibels in air, which is louder than a 747 at take-off.

If we follow the DC model and allow our Aquaman to talk to sea life, and communicate telepathically with ocean creatures, the sound of the sea will continue in this brutal vein. As our strategy Aquaman feeds his blubbering frame, consuming countless thousands of creatures to keep from freezing, he can hear the psychic screams of every zooplankter he gulps down his gullet.

His prey aren't the only ruckus in his soundscape. Throughout the ocean, predators chomp on the tongues, eyes, and gonads of their terrified victims. As Aquaman is programmed to adore the ocean with every bone in his body, the cacophony of dying sea creatures will surely break his heart, as *Aquaman extinct!* once more flashes green across your screen. And we haven't even begun to consider the atrocities heaped upon the sea by human beings. Aquaman senses the scars left by every trawl, knows personally the identities of many a fish and dolphin, casually murdered by our nets and lines. This strategy gameplay of Aquaman has quickly become the endless, unbearable drone of unrelenting murder. Cue end-game scenario, where Aquaman drinks like a fish as a desperate coping mechanism of life under the waves.

DID SUPERMAN REALLY TAMPER WITH TIME?

Interior. Car—day

LOIS lies buried, dead.
Exterior. Desert road—day

Heart-broken, SUPERMAN lays the dead LOIS onto the ground. Stroking her face, SUPERMAN bends down to kiss her. He then lets go of the body completely, chastising himself as she almost falls from his grip. . . . He starts shaking his head.

SUPERMAN: "No. No. (Looks up at the sky). NO. NO!"

SUPERMAN screams his anguish, and, eyes shut, rockets up into the sky. . . .

Exterior. Sky—day

SUPERMAN flies up the cloud layers.

JOR-EL'S VOICE: "My son . . . It is forbidden for you to interfere in human history."

SUPERMAN makes up his mind and flies up and onward as lightning flashes again.

Exterior. Space—low earth orbit

SUPERMAN flies incredibly fast, tracing a route around the world.

Shot of SUPERMAN'S face as he flies, a mask of grim determination and pain. There are blue streaks all around him, possibly due to relativistic effects of light being blue-shifted at such high speed.

Exterior. Space—low earth orbit

At super speed, he makes multiple transits around the world and the Earth begins to slow down. The transits go faster and faster, we cannot keep count as the world finally stops rotating on its axis. But SUPERMAN keeps going, not letting up, and the Earth starts rotating backward.

Exterior. Desert—day

Rocks and earth slide up the hill, as history rewinds.

Insert shot of SUPERMAN'S face as he interferes with human history this way, still flying, still a mask of grim determination and pain. The last transit ends, and SUPERMAN streaks away a short distance from the planet. Then he turns around and starts flying around the world in the opposite direction, starting the Earth rotating on its axis properly and allowing history to resume once more.

Exterior. Car—day

LOIS tries to get her car started, many times. But she has no success at it before SUPERMAN floats down to the ground beside the vehicle. LOIS feels something new has happened, and looks around before spotting him outside. She rolls down the window.

LOIS (gesturing wildly while SUPERMAN grins behind her, unseen): "I had this gas station blow up beside my car, there were telephone poles falling all over the road, I'm almost killed, and to top the whole thing off this stupid car runs out of gas!" (Turns to face him)

SUPERMAN: (smiling) "Well, I'm sorry about all that Lois. But I've been kinda busy for a while."

—Mario Puzo, David Newman, Leslie Newman, and Robert Benton, *Superman* (1978)

"The Time Traveler proceeded, 'any real body must have extension in four directions: it must have length, breadth, thickness, and duration. But through a natural infirmity of the flesh, which I will explain to you in a moment, we incline to overlook this fact. There are really four dimensions, three which we call the three planes of space, and a fourth, time.'"

—H. G. Wells, *The Time Machine* (1895)

Tampering with time has long been a hit in superhero stories. After all, if you dream about having the strength of a thousand men, or about flying as fast as a Peregrine falcon, or summoning up the natural forces of the planet, tinkering with time is pretty much going to be close to the top of your shopping list. If time could be mastered, if its relentless flow could be stemmed, then time's brutal agency of decay and death could be cheated. It's the ultimate superhero challenge.

It was H. G. Wells in his 1895 novel, *The Time Machine*, who first suggested the cosmos has four dimensions. Wells's Time Traveler was among the first to point out that the first three dimensions are that of space, and that time is the fourth. And the Time Traveler goes on to say, "There is, however, a tendency to draw an unreal distinction between the former three dimensions and the latter, because it happens that our consciousness moves intermittently in one direction along the latter from the beginning to the end of our lives." Traveling in time became an imagined possibility.

Wells certainly wasn't the first to dream of travels in time. There had been earlier folklore tales of temporal flirtations, but back in that day, it was usually dreamy magic mixed up with myth. For example, Samuel Madden's 1733 story, *Memoirs of the Twentieth Century*, featured a guardian angel, which travels back to the year 1728 with letters from 1997 and 1998. Johan Herman Wessel's tale, *Anno 7603*, written in 1781, is a story about a good fairy that transports people to the year 7603, where they find a society in which gender roles are reversed. And, perhaps most famously, Washington Irving's 1819 story of *Rip Van Winkle*, in which a man merely falls asleep on a mountain and wakes up twenty years in the future. There's even time travel in Charles Dickens's famous story, *A Christmas Carol*, but again the heartless old miser Ebenezer Scrooge is transported through time by ghosts of Christmas past and future.

But H. G. Wells was among the first to suggest a *mechanized* notion of time travel, a time travel device, or machine, which related back to the ancient Greek conception of time itself. The original Greek idea of time was a double identity, *chronos* and *kairos*. *Kairos* suggested a moment of time, in which something special happens. *Chronos* was more focused on measured, sequential, and mechanical time. Science and technology brought the mechanistic approach to the fore. *Chronos* became king. And time travel was born.

Time was all the rage. Concepts of time were splashed upon the canvas of the Cubists. Painters like Picasso and Braque created artwork, in the earliest years of the twentieth century, where various viewpoints were visible in the same plane, at the same time. All four dimensions were used to give the viewer a more profound sense of depth. It was a revolutionary new way of looking at reality. Time was present in the stop-motion photography of Étienne-Jules Marey in the 1890s, and was soon captured in moving pictures. In 1912, it inspired French artist, Marcel Duchamp, to paint his highly controversial *Nude Descending a Staircase*, which depicted time and motion by successive superimposed images.

Space-time was born. Einstein gifted this already-buoyant culture a new perspective on the fourth dimension. His relativity theory conjured up some striking new ideas about the nature of time in the cosmos: moving clocks run slow; time is slowed down by gravity; and the speed of light is the same no matter how the observer is moving. It was a revolution in time. And it seemed to cause distress in Spanish surrealist painter Salvador Dali. For many, his anxiety is palpable in his famous 1931 painting, *The Persistence of Memory*. Dali's floppy clocks are history's most vivid depiction of Einsteinian gravity distorting time.

COMIC BOOKS CATCH UP WITH TIME

Comic books took their time to catch on, but soon time travel stories surfaced in comics too. In 1928 came a story called, *Armageddon 2419 AD*. In this tale, a strange cave gas transports Anthony Rogers to the year in question. The story was the basis of the now-familiar comic-strip, *Buck Rogers*. In 1942, during World War II, *All-Star Comics #10* featured a story in which the *Justice Society of America* travel five hundred years into the future to secure an effective defense against bomb attacks. And just after the war had ended, and hot on the heels of Hitler, a supervillain murders a scientist, who has invented a time machine, and tries to use the time device to alter history so modern technology cannot defeat his bid to conquer America. Finally, in 1952, Ray Bradbury's story, *A Sound of Thunder*, was one of the tales that marked the beginning of the fashion for stories that create alternate timelines and plots. Bradbury's "butterfly effect" meant changes made in the past hugely affected the fictional future.

DC Universe's prime time traveler was The Flash. His ability to tear around at super speeds meant he could combat enemies not merely in the comic book present, but also in the far future of the thirtieth century. The Flash was just one example of the new trend in time-twisting tales, which became more and more complex. And so time travel, which began as a relatively simple concept, ended up with worlds and timelines co-existing in a multiverse so complicated that DC had to wipe the time-travel slate clean and start again.

Perhaps the best and most famous example of a DC superhero getting caught up in the complexity of time travel is that of Superman. In the 1978 movie, *Superman*, our eponymous hero Kal-El is at first unable to prevent the death of Lois Lane. In a rage, and unable to passively accept his loss of Lois, Superman decides to fly above planet Earth, his adopted home. He purposefully flies around the Earth at such speed that, allegedly, the world starts spinning backwards. Time is rewound and, presto, the world goes back to normal and Lois is re-spawned, as if reality were all a kind of giant video game. But would this really have worked? Is it a method of time tampering that might produce Superman's desired result of rewinding time?

REPLAYING THE MOVIE

To make sure we know what Superman is meant to be doing, let's replay the scene. When you first see Superman's illuminated light trail zooming about the girth of the Earth, you might think that he's meant to be flying faster than light. But, on further examination, it becomes clear that he's meant to be traveling in time with the aim of making the Earth spin in reverse. Then, once he's made the world spin backwards, he decelerates and somehow manages to spin the world the other way again. And so it seems we are expected to believe that Kal-El is altering the Earth's rotation, supposedly using some kind of force, and that the very rotation of the Earth is what determines the flow of time in the entire cosmos, as if the

Earth was some kind of Aristotelian metronome, about which the rest of the Universe ticked.

You may not be too surprised to find that forcibly altering the spin of the Earth won't actually rewind time. The rotation of the Earth is *not* what governs the flow of cosmic time. But, fact check number two: Is Superman even traveling faster than light speed anyhow? If we're pedantic enough to painstakingly count his revolutions of the Earth, we'll find that Superman makes about sixty-five trips about the globe. The time taken for these trips is just over 17 seconds. And, given that the Earth has a circumference at the equator of 40,000 kilometers, we find that Superman is in fact only traveling at *half* light speed. So, even if it were possible to hit light speed and reverse the Earth's rotation, Superman simply isn't going fast enough.

CAREFUL WITH PLANET EARTH, KAL-EL

But how would he reverse the Earth's rotation? To stop the planet from spinning, Superman would either have to grapple directly with the Earth in some kind of way, or else slow it down, changing its spin by an indirect drag force. Drag, sometimes known as air resistance, is a type of friction that applies to moving bodies in fluids, such as water and air. A drag force acts in an opposite way to the relative motion of an object moving through the fluid. As drag depends partly on the density of the fluid surrounding the object in motion, the drag force behind the speeding Superman would transfer to the surrounding air and some might even make it down to the surface of the Earth, acting as a brake on its rotation. But the air drag behind Superman would have to do a huge amount of work to slow the Earth. That's because the Earth has an incredible amount of rotational inertia. Inertia is the force that keeps an object moving in the same direction, and it's a resistance to change, so in this case the Earth would be resistant to having its spin altered. In fact, the Earth has a mass of 5.9736×10^{24} kilograms, or roughly 5.9 sextillion tons, and is spinning at around a thousand miles an hour. And this data means that the energy needed to slow down the planet, let alone stop it dead and then spin it back in the opposite direction, would be the same amount of energy that would not only destroy the Earth, but also separate it gravitationally, as if it had been zapped by the Death Star (please see my *Science of Star Wars* book!)

It's just as well Superman wouldn't be able to slow the planet and stop it. The consequences would be quite catastrophic. It'd be akin to having a ton of water in the back of a flat-bed pickup, driving at breakneck speed, and then suddenly stopping somewhere on the highway. Now imagine all the Earth's oceans spinning at a thousand miles an hour, and then just stopping. It'd make the tsunami on the movie *San Andreas* look like a ripple in the bath. Incidentally, since we're on the topic of water and the spin of the Earth, humans today control water on a massive scale. Our reservoirs hold over 10,000 cubic kilometers of water. That's five times the amount of water in all the rivers on Earth. And as most of it is pooled in the more populated northern half of the planet, the extra weight of the water has changed how the Earth spins on its axis. It's caused the Earth's rotation to speed up, shortening the day by 8 millionths of a second in the last forty years.

One final word on traveling at light speed. If Superman flew above the planet's surface with increasing acceleration, innocent bystanders would first hear him breaking the sound barrier. Then, as he sped up and reached speeds equivalent to spacecraft reentering the Earth's atmosphere, the air in front of him is heating up to such an extent that it begins to transform into a cloud of glowing plasma, forming a shroud around Superman as he rockets past. The shockwave that comes with this Superman shroud would radiate down to the ground and help annihilate people under Kal-El's flight path. Finally, once he'd reached half-light speed, the air ahead of him undergoes nuclear fusion, releasing gamma rays, and the plasma around him releases x-rays, and the accompanying shock wave so that Superman becomes a cone of nuclear annihilation wherever he flies.

DAILY DIARY: LIVING LIKE LOKI—IS IMMORTALITY A DRAG?

"Personally, I would not care for immortality in the least. There is nothing better than oblivion, since in oblivion there is no wish unfulfilled. We had it before we were born yet did not complain. Shall we whine because we know it will return? It is Elysium enough for me, at any rate."

—HP Lovecraft, *Selected Letters V* (1934–1937)

When all the world was very young
And mountain magic heavy hung
The supermen would walk in file
Guardians of a loveless isle
And gloomy browed with super fear their tragic endless lives
Heaves nor sighs in solemn, perverse serenity
wondrous beings chained to life.
Strange games they would play then
No death for the perfect men
Life rolls into one for them
So softly a super god cries

—David Bowie, *The Supermen*, on *The Man Who Sold the World* (1970)

"I do not believe in immortality of the individual, and I consider ethics to be an exclusively human concern with no superhuman authority behind it."

—Albert Einstein, *Albert Einstein: The Human Side* (1981)

"I don't want to achieve immortality through my work. I want to achieve it through not dying."

—Woody Allen, *On Being Funny* (1975)

He was beautiful. It was a beauty that was hard to fix, or to see. For he was at all times shifting, shimmering, fluttering, melting, and mixing. He was the form of a shapeless flame. He was the turbulent thread of molecules in the flowing mass of the waterfall. He was the nameless wind that blew the clouds in billows and bands. One might spy the barest of trees on the skyline bent by the wind, reaching out its twisted roots and branches, and swiftly its shape would suddenly resolve into that of Loki, the trickster.

Loki was half way to immortality. As someone from the race of Frost Giants of Jotunheim, he commanded a huge range of superhuman traits in practically all activities, but key characteristics here was a durability enough to withstand high-caliber bullets, and an immunity to all known diseases and toxins, as well as resistance to aging.

Immortality is one of the essential themes of comic book classics, and of speculative thought in general. Immortality is usually wrapped up with notions of the elixir of life, and the fountain of youth, both of which are common hypothetical goals of various quests within science fiction. And what this usually means, as in the case of Loki, is extreme longevity, freedom from ageing, and relative indestructability.

But fiction writers and thinkers have always had doubts about immortality. It's often treated as a false aim, echoing the perpetual punishment handed out to the likes of Sisyphus, the king of Greek mythology who was punished by being forced to forever roll an immense boulder up a hill, or the Wandering Jew, who taunted Jesus on the way to the cross and was then cursed to walk the Earth until the Second Coming.

Writers have rarely been keen on the social impact of immortality. Many have suggested immortality would lead to a kind of social sterility—a society going nowhere, and subject to little, if any, change. Even those writers who took a brighter view chose a rather elitist conclusion of a few privileged immortals living in a world of mortals. And when George Bernard Shaw showed enthusiasm for universal longevity in his 1921 work, *Back to Methuselah*, Karel Capek, the sci-fi writer who introduced the word "robot," responded to Shaw by saying immortality would be an unmitigated nightmare, even for a single person.

So, you may well ask, what's so bad about living forever? To a five-year-old boy, one year is 20 percent of his life. To his twenty-five-year-old mother, one year is only 4 percent of her life. The same 365 days feels very differently to different people. With modern advances in medicine, it's quite possible the boy could live for a hundred years, or 36,500 days. But imagine living for 36,500 years.

If our boy lives for 36,500 years, then a single year could feel to him like a day. Let's go back to Loki. Would the boy's emotions hold true through the likely boredom of living for millions of years? Maybe this is why Loki seems so sociopathic. Humans might become very sad and lonely, knowing they have and forever will outlive everyone they have ever loved. But does Loki care about that? Perhaps he is the perfect immortal, and the constant change in movement and mood is down to sheer boredom.

But what if everyone were immortal? Research into biotechnology following the cracking of the genetic code has meant science really may be able to keep the spark of life alive. The recipe for doing this a varied one, running from treatments and medicines, such as eugenics and genetic engineering, all the way through to artificially extending life, with the use of synthetic organs or by becoming a cyborg. Another option for immortality involves uploading human consciousness into new bodies. In one sci-fi story that considered this immortal future, the humans did this uploading so many times that they begin to see themselves as gods. The slippery slope to becoming a sociopath like Loki!

A PLANET EARTH, REPLETE WITH IMMORTALS

Imagine an Earth replete with such immortals. Our planet is only so big. Where would we all live? This kind of question was tackled in the 1966 sci-fi novel, *Make room! Make room!* Written by American novelist, Harry Harrison, the book explored the consequences of a huge population growth on society. Set in a future where the global population is seven billion (uh oh), the world suffers from overcrowding, resource shortages, and a crumbling infrastructure. The novel also acted as the basis for the 1973 movie, *Soylent Green*, where the film went one better (or worse!) by introducing cannibalism as a way to feeding the growing number of people.

And what about the immortal darker side of dating? Imagine our immortal boy finds falls in love with a companion once every hundred years.

That means our serial monogamist immortal would have ten thousand girlfriends in a million years. That's some challenge. And it would totally change the definition of what a meaningful relationship means. For one thing, how many of those ten thousand companions' names will he be able to remember?

In fact, memory would be a problem for immortals. People can rarely remember in detail what they did last year, and especially when they were five years old. Think about it: What proportion of your past have you forgotten? Unless you have a truly exceptional memory, probably the vast majority of your past is lost forever. And if we have problems recalling what we did when we were five, how much would we remember if we lived for a thousand years, or if we lived to a million years old? Human brain capacity is limited. And that means we simply don't recall all of life's minutiae, as our minds merely replace useless memories, such as answers to the trivial security questions banks always expect you to remember, with far more vital data!

Another dastardly dynamic with immortality is all down to Darwin. It's this: people have not always looked the same. According to the Darwin-Wallace theory of evolution, as women find taller men more attractive, then taller men would be more likely to mate and have kids. And that means more tall genes in the gene pool so, in the next generation, more kids will have the genes to be taller. Repeated iterations of this sexual selection over one million years would mean that average human height will be lots taller than today's average height. Assuming, of course, that humans aren't completely wiped out by natural disasters, such as an asteroid impact, a giant solar flare, or a global epidemic.

By most accounts, human ancestors were short, hairy apes. Now, we don't look like apes any more. But, although we could be called "naked apes" compared to our closest living relatives, we actually have as many hairs per square inch on our bodies as there are on a chimpanzee. The human body is completely covered in hair, from head to toe. But, as many of those hairs are so fine and fair, they're just not visible to the human eye. And we haven't always looked this way, of course. Our appearance has slowly evolved and gradually thinned our body hair over time. Now, imagine you are the only immortal, living in a world of mortals. Everyone would continue evolving, generation after generation, so that you will end up looking very different from everyone around you. After all, imagine we pulled off a Jurassic Park trick on a Neanderthal. Most modern humans would fail to make friends with the fella, and simply call the local Museum of Natural History.

And one more drawback of being an immortal like Loki: polytrauma. Immortality is one thing. But invincibility is quite another. And being immortal doesn't necessarily mean being invincible. It merely means you can't die. Immortality doesn't provide you with a warranty for body health. Think about a normal human body and how many scars it might have. And then think about how many permanent scars an immortal will have after living for just a thousand years!

Consider mutilation in the United States. In one year alone, there are almost one-fifth of a million amputation-related hospital discharges in the US, mostly due to illness or accident. That percentage seems low when you compare it to a total population of 325 million souls and a life expectancy of around eighty years. But if you've been alive for a million years, the chance of you keeping all your appendages is woefully slim. In fact, you're more likely to be a victim of major trauma, or polytrauma, and your injury severity score (ISS), a kind of medical score given to assess trauma severity and occurrence, would be high.

And think about those exquisite bodily essentials, such as eyes, nose, teeth, and toes. What are the chances of you retaining all your teeth, or even both your eyes, for a thousand, or especially a million years? You're far more likely to look less like Loki and more like Two-Face, Deadpool, Shithead, or Yellow Bastard. Are you still sure you want to live as long as Loki?

CAPTAIN AMERICA: EXPERIMENTS WITH ÜBERMENSCH

"I teach you the overman. Man is something that shall be overcome. What have you done to overcome him? . . . All beings so far have created something beyond themselves; and do you want to be the ebb of this great flood, and even go back to the beasts rather than overcome man? What is ape to man? A laughing stock or painful embarrassment. And man shall be that to overman: a laughingstock or painful embarrassment. You have made your way from worm to man, and much in you is still worm. Once you were apes, and even now, too, man is more ape than any ape . . . The overman is the meaning of the earth. Let your will say: the overman shall be the meaning of the earth . . . Man is a rope, tied between beast and overman—a rope over an abyss . . . what is great in man is that he is a bridge and not an end."

—Friedrich Nietzsche, *Thus Spoke Zarathustra* (1883)

"I do believe that man is a rope between animal and superman. But the superman I'm thinking of isn't Nietzsche's. The real superhuman, man or woman, is the person who's rid himself of all prejudices, neuroses, and psychoses, who realizes his full potential as a human being, who acts naturally on the basis of gentleness, compassion, and love, who thinks for himself and refuses to follow the herd. That's the genuine dyed-in-the-wool superman."

—Philip Jose Farmer, *The Dark Design* (1977)

"I hate the whole übermensch, superman temptation that pervades science fiction. I believe no protagonist should be so competent, so awe-inspiring, that a committee of twenty really hard-working, intelligent people couldn't do the same thing."

—David Brin, *Locus* (1977)

Why does science fiction bother itself so with supermen?

Take Captain America, for example. Steve Rogers had been a scrawny fine arts student growing up during the Great Depression. Understandably appalled at the horrific atrocities of the Nazis, Rogers tried to enlist in the army, but his puny human frame ended in "#fail." Up steps Professor Abraham Erskine, a typical science-fictional scientist with a superman obsession. Erskine's pet project was to use bleeding-edge science to enhance US soldiers to the height of physical perfection. After injections and ingestion of a "Super Soldier Serum," Rogers became the sole beneficiary of Erskine's genius, altering his physiology from a frail #superfail to a maxed out #superman.

This is the weird thing about science fiction. Most mainstream fiction since the Renaissance hasn't bothered itself with the world revealed by science and wonder. Poetry had little to do with the laws of physics, that was the chief idea. But there were some early pioneers, the early nineteenth century Romantic poets, for example, who became enchanted with the new and evolving science of the day. Witness William Wordsworth's words, from his *Lyrical Ballads* of 1798, "If the labors of men of science should ever create any material revolution . . . in our condition . . . the poet will sleep then no more than at present, but he will be ready to follow the steps of the man of science, not only in those general indirect effects, but he will be at his side, carrying sensation into the midst of the objects of the science itself."

Trying to best express, "the taste, the feel, the human meaning of scientific discoveries" is an early expression of how science fiction works. Science fiction is all about science and progress. It's a mode of thinking, whose main obsession is closing the gap between the new worlds uncovered by science, and the fantastic strange worlds of the imagination. And that includes the possibility of becoming supermen. Back in those days of the Romantics another obsession started. Evolution.

The word is synonymous with Charles Darwin, naturally, but an early and provocative evolutionist was Charles's grandfather, Erasmus. As a boy Charles had poured over Erasmus's mighty work on evolution, *Zoomania*, published in two volumes in 1794 and 1796. It was replete with hearty exclamations that life had evolved from a single ancestor.

Romantic poet Samuel Taylor Coleridge declared Erasmus Darwin "the first literary character of Europe, and the most original-minded man." One of Erasmus's poems on evolution enjoys a science fictional vision. It foresees, with unerring accuracy, a future of colossal skyscraper cities, overpopulation, convoys of nuclear submarines, and the advent of the car. It's easy to glimpse the early links between the Romantics, science fiction, and supermen.

When Charles published his own book on evolution, the famous 1859 work, *The Origin of Species,* so began another new paradigm: the process of becoming; the question as to what would become of man. In the words of Alfred Lord Tennyson, "Earth's pale history runs, what is it all, but a trouble of ants in the gleam of a million, million suns?" This irresistible rise of the metaphor of evolution spawned around seventy futuristic fantasies in England alone between 1870 and 1900.

But the work that was to make its mark most on the question of supermen was more fact than fiction. And that work was *Also Sprach Zarathustra*, written by German philosopher, Friedrich Nietzsche, and inspired in turn by Darwin. Nietzsche's book identified three stages in the evolution of man: ape, modern man, and ultimately, superman. As Nietzsche put it, "What is the ape to man? A laughingstock, or painful embarrassment. And man shall be to the superman: a laughingstock or a painful embarrassment." Modern man is merely a link between ape and superman. For the superman to evolve, man's will, "a will to procreate, or a drive to an end, to something higher and farther," must power the change.

2001: A SPACE SUPERMAN

A good example of how Nietzsche's work fed into fiction is the 1968 movie, *2001: A Space Odyssey*, by Arthur C. Clarke and Stanley Kubrick. The film was delivered during the infamous extraterrestrial hypothesis, which peaked between 1966 and 1969. The hypothesis held that UFOs were close encounters with visiting aliens, a hypothesis vastly influenced by science fiction, of course. Celebrated for the maturity of its portrayal of mysterious, existential, and elusive aliens, *2001* had raised science fiction cinema to a new level. Eminent US film critic Roger Ebert, when asked which films would remain familiar to audiences two hundred years from now, selected *2001*. Another critic claimed the picture was an, "epochal achievement of cinema" and "a technical masterpiece." The film, not the book, made Arthur C. Clarke the most popular science fiction writer in the world. And Kubrick's masterpiece quickly became a classic discussed by many, if not understood by all.

For, like Captain America, *2001: A Space Odyssey* is all about supermen. Like Nietzsche's work, Kubrick's motion picture traces man's journey through three stages. As the movie's subtitle suggests, the narrative is a spatial odyssey from the subhuman ape to the post-human starchild. The unfolding four-million-year filmic story embraces each theme of science fiction: space (contact through alien cultural artifacts), time (evolutionary fable), machine (the man-machine encounter with HAL, computer turned murderer), and monster (human metamorphosis). The opening "Dawn of Man" scene of *2001* sees the Sun rise above the primeval plains of Earth, to the rising soundtrack of Richard Strauss' Nietzsche-inspired tone poem, *Also Sprach Zarathustra*. A small band of man-apes are on the long, pathetic road to racial extinction.

But this is a Darwinian tale about transcendence, not extinction. The journey to superman begins with one of the hominids proudly hurling an animal bone into the air. In an astounding cinematic ellipsis, the bone instantly morphs into an orbiting satellite, and three million years of hominid evolution is written off in one frame of film. So much for man. The agency that drives the guided evolution of these early hominids is an alien artifact in the shape of a black monolith. Primal bone technology marks the birth of the modern era. Man and machine, from the very outset, are inseparable. The mysterious presence of the monolith transforms the hominid horizon. The journey to superman begins.

CAPTAIN AMERICA: NIETZSCHEAN ÜBERMENSCH

The influence of Nietzsche also went into the creation of Captain America. The first Captain America comic, cover-dated March 1941, but on sale December 20, 1940, showed the protagonist punching Nazi leader Adolf

Hitler. The comic came out a year before the attack on Pearl Harbor, but more than a year into World War II. It sold around one million copies. The coupling of the Captain and Hitler was a curious one. Both had been influenced by Nietzschean philosophy.

In his prologue to *Also Spake Zarathustra,* Nietzsche had called his Übermensch or "superman" the savior of humanity. In contrast, he talked of the "Last Man," (German: *der letzte Mensch*) or the "Last Race." The Last Men are the antithesis of his imagined superior being, the Übermensch. The Last Men, says Nietzsche, are tired of life. They take no risks. They seek only security and comfort. Nietzsche derides modern society and Western civilization. He says that the Last Race is the goal they have apparently set themselves. The lives of the Last Men are pacifist and contented. There is no distinction between ruler and ruled. No division of strong and weak. No dominion of supreme over mediocre. Social struggle and challenges are marginalized. Individuals live equally, and in "superficial" harmony. There are no novel or thriving social trends and ideas. Individuality and creativity are suppressed. Zarathustra tries but fails to get the populace to accept the Übermensch as the preferred goal of society. Instead, they choose the "disgusting" goal of becoming the Last Men, which appalls Zarathustra.

Nietzsche warned that the Last Race was doomed to fail. The society of the Last Man would be too sterile and degenerate to foster the growth of healthy human life, and especially great people. In Nietzsche's view, the Last Man was only made possible by mankind having bred inferior, apathetic people, or ethnic groups unable to dream, unwilling to take risks, reducing their being to mere existence, earning their living and keeping warm. The society of the Last Man is opposite to Nietzsche's notion of the human will-to-power, the main driving force and ambition behind human nature, along with all other life in the universe.

The rhetoric of Nietzsche's Übermensch was famously used by Hitler. His German fascist regime was eager to lay claim to Nietzsche's ideas, to use the ideas to justify Nazi atrocities during World War II, and to portray Nazis as inspired by Nietzsche. In 1932, Nietzsche's sister received a bouquet of roses from Adolf Hitler, and in 1934 Hitler personally presented her with a wreath for Nietzsche's grave with the words, "To A

Great Fighter." Again in 1934, Nietzsche's sister gave Hitler her brother's favorite walking stick, and Hitler was pictured gazing lovingly into the eyes of a white marble bust of Nietzsche.

Nietzsche's take on Darwinian ideology was also among the triggers for the eugenics movement in the United States.

It was the cousin of Charles Darwin, Francis Galton, who had coined the term eugenics. Galton claimed to have traced what he saw as superior human qualities being passed down from generation to generation among Europe's most eminent men. In contrast, he suggested that weak, inferior, and even dangerous traits were also being passed down—most clearly, in Galton's eyes, in society's lower classes, and within certain races. Galton believed in the inequality of humans. For example, he thought Africans were inferior and suggested that the east coast of Africa be settled by the Chinese who were, according to Galton, superior. Galton's eugenics plan was twofold. Firstly, he proposed a human breeding program to produce superior people. The "Super Soldier Serum" administered to Steve Rogers by Professor Abraham Erskine immediately springs to mind when reading of Galton's first plan. And Galton's second plan was to improve the quality of the human race by eliminating, or excluding, biologically inferior people from the breeding population.

Such forced sterilization was practiced in the US. It was used as a way of trying to control so-called "undesirable" populations—the poor, immigrants, people of color, disabled people, unmarried mothers, and the mentally ill. Federally-funded sterilization programs were carried out in thirty-two states, for most of the twentieth century. The programs were greatly influenced by Nietzsche's and Galton's prejudiced ideas of science and social control. Alex Stern, author of a book entitled *Eugenic Nation: Faults and Frontiers of Better Breeding in America*, exposes the sheer scale of the program, "In the early twentieth century across the country, medical superintendents, legislators, and social reformers affiliated with an emerging eugenics movement joined forces to put sterilization laws on the books. Such legislation was motivated by crude theories of human heredity that posited the wholesale inheritance of traits associated with a panoply of feared conditions such as criminality, feeblemindedness, and sexual deviance. Many sterilization advocates viewed reproductive surgery as a necessary public health intervention that would protect society from deleterious genes and the social and economic costs of managing 'degenerate stock.'"

California led the way. And the state's eugenics program even inspired the Nazis. Andrea Estrada at UC Santa Barbara, tells in *The UC Santa Barbara Current* how the forced sterilization was particularly rampant in California. "Beginning in 1909 and continuing for 70 years, California led the country in the number of sterilization procedures performed on men and women, often without their full knowledge and consent. Approximately 20,000 sterilizations took place in state institutions, comprising one-third of the total number performed in the 32 states where such action was legal." And the *LA Times* has quoted Hitler's admiration of these eugenic efforts. "There is today one state," wrote Hitler, "in which at least weak beginnings toward a better conception [of citizenship] are noticeable. Of course, it is not our model German Republic, but the United States."

Hardly surprising, then, that all this eugenic talk found its way into the creation of Captain America. Nietzsche's Übermensch philosophy is writ large in the backstory of the Captain. Steve Rogers, "scrawny, scary, starved," is a Last Man. Science not only "saves" him, but with an injection of a little super serum and genetic tinkering, Steve becomes the super soldier. The Captain's original role was as the nation's savior from internal spooks and saboteurs. And to play that role, Steve had to be saved from himself, so he can save America, wiping out all those "unfit" and "un-American." Remember that in *Captain America: The Winter Soldier*, Samuel L. Jackson, gifted a more prominent role as Nick Fury, says "You guys did some nasty stuff," in relation to the idea of America's "Greatest Generation," and the assumed American moral superiority over the rest of the world during WWII and beyond.

It was only later that the Captain's image was cast through a less illiberal lens. After the influence of the swinging sixties, the summer of love, flower power, and the civil rights movements across the West, the Captain was recast as the free world's first defense against Fascism. Rather ironic, given his Nietzschean origins. It's the liberal Captain we get in the cinema, a little less of the problematic Übermensch, and a little more egalitarian. In short, a far more nuanced character.

THE TROUBLE WITH ÜBERMENSCH

And yet, a problem remains. Not just with the Captain, but with all Übermenschen. Consider the case of George Orwell's novel, *Nineteen Eighty-Four*. No other novel written in the twentieth century has quite captured the popular imagination as much as Orwell's haunting specter of big government gone mad with lust for power. The very title of this classic dystopia became a cultural watchword. And the word "Orwellian" still ominously speaks of matters hostile to a free society. No single work of science fiction has had a greater impact on politics. Big Brother is almost as famous as Frankenstein. He and the concepts of Room 101, Newspeak, and the Thought Police are still with us in these days of euphemism, political bullshit, and fake news.

But *Nineteen Eighty-Four* is a flawed masterpiece. Orwell's book was used with little regard to the author's intention. And many critics suggest that this was made far easier by the book's unrelenting portrayal of defeat. From reading Orwell's work, few would feel they had a ghost of a chance of winning when pitted against big government. The novel reinforces passivity rather than undermines it. Likewise, narratives about Captains, X-Men, and superheroes (whether comic-book or cinema) place the audience in the same boat as readers of *Nineteen Eighty-Four*.

Watching the Captain act out as Übermensch relegates the viewer to the role of the Last Race. Nietzsche's Übermensch has isolated himself from the rest of humanity. Captain America is such a superhero, ascended way beyond the ordinary. Superior in all matters, he separates himself, as he no longer needs other people. So much so that he can be plunged into ice and pulled out seventy years later, removed from all he knew and all those who knew him. Like Orwell's *Nineteen Eighty-Four*, superhero stories reinforce passivity, rather than undermine it. They help make Last Men of us all. Writer James Gunn makes a good point towards the end of *Guardians of the Galaxy Vol. 2*. In an exchange where Peter Quill is about to smite his god-like Celestial father, Ego says to Quill, "WAIT! You're a god. If you kill me, you'll be just like everyone else!" To which Quill replies, "And what's wrong with that?!" And as Orwell said about the science fictional narrative of *Nineteen Eighty-Four*, "The moral to be drawn from [dangerous nightmare situations] is a simple one: don't let it happen. It depends on you." It is the Last Men that will inherit the Earth, not the Übermensch.

DAILY DIARY: HOW HAS EVOLUTION MADE REAL-LIFE X-MEN MUTANTS?

"Mutation: it is the key to our evolution. It has enabled us to evolve from a single-celled organism into the dominant species on the planet. This process is slow, and normally taking thousands and thousands of years. But every few hundred millennia, evolution leaps forward."

—Professor Charles Xavier in *X-Men*, David Hayter screenplay (2000)

"Natural selection is not fully sufficient to explain evolutionary change for two major reasons. First, many other causes are powerful, particularly at levels of biological organization both above and below the traditional Darwinian focus on organisms and their struggles for reproductive success. At the lowest level of substitution in individual base pairs of DNA, change is often effectively neutral and therefore random. At higher levels, involving entire species or faunas, punctuated equilibrium can produce evolutionary trends by selection of species based on their rates of origin and extirpation, whereas mass extinctions wipe out substantial parts of biotas for reasons unrelated to adaptive struggles of constituent species in 'normal' times between such events."

—Stephen Jay Gould, "The Evolution of Life on the Earth," *Scientific American Special Issue* (1994)

"The experimental detonation of a gamma bomb bathed him in massive amounts of radiation, enhancing his body."

"A mutant healing factor regenerates damaged tissue and makes him extraordinarily resistant to diseases, drugs, and toxins."

"A former test subject of the Weapon X program, possessing a regenerative healing factor and an unstable compiling of other Weapon X experiments."

"Radioactive mutagenic enzymes in the venom quickly caused numerous body-wide changes, primarily superhuman strength, reflexes, balance, and the ability to cling tenaciously to most surfaces."

—The mutable origins of superheroes Hulk, Wolverine, Deadpool, and Spiderman, respectively

Good old Charlie Darwin. If it wasn't for Charlie, we'd never be able to fathom how some of our favorite superheroes were formed. Consider Storm, for example. Also known as Ororo Munroe, Storm is muta-genetically imbued with the potent ability to conjure the elemental forces that govern the weather, as well as other atmospheric phenomena. This would be particularly useful in countries like Darwin's homeland of Britain, where it seems to rain incessantly. Then there's Jean Grey, one of the five original X-Men. Jean is an omega-level mutant telekinetic and telepath with almost limitless powers, which have seen her handily return from death many times. It's not known if Jean also has a genetic way of avoiding premature funeral costs. The ultimate mutant weapon is Wolverine. With the dramatic refrain of, "with the rage of a beast and the soul of a Samurai," James "Logan" Howlett has a mutant way of living a long life through an accelerated healing factor and enhanced senses, though you do wonder if he's ever able to totally relax when he's picking his nose.

Then there's Professor Charles Xavier. Professor X (simply a superb sounding comic book name, even better than Doctor Strange) is the creator of the X-Men and founder of the Xavier School for Gifted Youngsters. Professor X is the true brains behind the mutants, and he dreams of peaceful coexistence between mutants and humans. His mission is helped by his mutations: he's a hugely powerful telepath and scientific genius. Being among the world's greatest masterminds means that Professor X knows all about Darwin and the origins of the idea of evolution. And he knows the story of evolution started not with mutations, but with rocks.

DARWIN AND THE ROCKS

Professor X would be well aware of the fact that Charlie Darwin is said to have started the ball rolling on the understanding of evolution. But, the truth is, of course, Professor X would know Charlie had help. The theory of evolution was not the work of a single "genius." Since antiquity, some of the finest philosophers have wondered about the rich variety of life. During that time, divine creation had not always been thought of as the causal factor. A stream of very able thinkers—some even brainier than Professor X himself—such as Greek thinkers Empedocles, Epicurus, and Lucretius, through to Renaissance Italian Leonardo da Vinci, had tended towards a more secular speculation. Instead of believing in the Great Chain of Being, that every form of life was created by God and remained unchanged since, these sages looked to nature's inherent patterning for a cause. And in the days of Darwin himself, Britain could brag of another brilliant biologist. Younger than Darwin by a decade, the naturalist Alfred Russel Wallace was radical and open-minded, an autodidact who was both intuitive and unconventional. And his work helped Darwin immensely.

In the view of many scholars like Professor X, the story really starts during the Industrial Revolution, when great engines turned over the soil of the world. The steam engine helped open up the veins of the Earth. And in developing industrial nations, especially Britain and Germany, scholars began to learn how to read the rocks. They learned that the same strata were found always in the same order, and contained the same fossils. And that meant that the Earth's history could be read from the fossil sequence, contained within the rocks.

Rockreading revolutionized the world. Scholars soon learnt to read and understand the rocks, just like they would a book. And the fossil record also began to churn out the signatures of huge beasts that no longer stalked the Earth. The discovery of dinosaurs and other creatures led scholars to a startling conclusion. Planet Earth must be truly ancient to account for all the changes found in the fossil record. Slow, but unstoppable, change in the workings of the Earth must mean the world's history is huge. And the ruins of an older world must be visible somewhere in the structure of our planet. All the scholars need do is dig.

And boy did they dig. What they found in the depths of the Earth was truly remarkable and enough to make the heart of Professor X sing. They unearthed a cornucopia of mutant plants and creatures. Jaw-dropping fossilized flowers that no one had ever seen in bloom. Yomping woolly mammoths that looked like elephants in need of a haircut. And the "terrible lizards," those huge dinosaurs, along with a whole host of bizarre beasts that looked like they had all just walked out of someone's dream, or nightmare. The fossils were evidence that the Earth had seen great change in its long history. Paleontology, the study of fossil reading, was soon founded by Baron Georges Cuvier, also known as the "Pope of Bones."

And this study of the fossil record meant that some scholars came to a revolutionary conclusion about the secrets of the Earth: the theory of evolution. The procession of the animals and plants found in the Earth's rocks meant only one thing. That all living things changed from earlier times, during the history of our planet. Evolution also says that all living things are related. So, if you go back far enough in time (far enough down into the rocks of the Earth), all animals, all plants, and all living things, had one ancestor. And, as Professor X might conclude, evolution led to that one species becoming so many more, until today where we have millions upon millions, humans and mutants included.

Now, Professor X would also be aware of the fact that the theory of evolution is like a ticking time bomb. Yes, evolution means we share a history with all the animals on the planet today, and all the extinct animals found in the ground. But that fact requires millions and millions of years for the slow changes in the planet to happen. So, if the Earth's history *is* huge, and all the fossils are evidence of older times, scholars needed a way to date the rocks. The dating helped prove that evolution was a very viable theory. They needed to tell how long ago layers of sediment were set down to become strata, deep in the Earth. Luckily, buried deep in the Earth's rocks are many chemical elements. And some of these elements are radioactive, which means they decay from one type of chemical into another. The amount of time it takes for exactly half of one element to decay into another is known as the half-life. So scholars can measure how much decay has happened and, using the half-life, work out how long the element has been decaying in the rock. It's sort of like a ticking clock. And it means scholars can date the rocks and measure evolutionary time.

MUTANT TIMES

And yet, in those days of Darwin, only part of the story of evolution was known. Scholars think they understood evolution by natural selection. This was the natural process that results in the continued existence of only the types of plants and animals that are best able to produce young, or new plants, in the conditions in which they live. This is how most people think of evolution—the adaptive struggle of species to survive in a given environment. But as Stephen Jay Gould points out in his essay, "The Evolution of Life on the Earth," in a *Scientific American* special edition, there is evolution and biological organization both above and below this traditional Darwinian focus. At the higher levels, entire species might get dumped into the dustbin of history by a mass extinction event, such as the kind of cosmic catastrophe that is favored in the pages of comic books, like an asteroid or comet collision with the Earth. And at the lowest level, evolutionary change happens through mutation.

To get a grip on mutation, imagine Professor X standing in front of a mirror. Professor X is looking at a human body, made up of trillions of cells, the basic unit of all living things. There are many different types of cells, and each does a different job. But inside, almost every one of Professor X's cells is the same. Each cell has a nucleus, which contains 99.9 percent of his genes. And cells also have mitochondria, which contain a few more genes. Altogether, Professor X's body has around twenty thousand genes.

As you may know, Professor X's genes are a small part of a chemical called DNA. DNA is a double-stranded chemical, made up of sugar, phosphate, and four different bases: adenine, thymine, cytosine, and guanine. There are nearly two meters of DNA squeezed into almost every human cell. But the genes are scrunched up so tightly that they're able to squeeze into just one cell nucleus. The four bases in DNA spell out the language that scholars the world over call the genetic code. The number and order of the bases decide, for example, whether you are Professor X, a banana, a chimpanzee, a cow, or a normal human. Most genes are recipes for making particular proteins. And it's these recipes that are passed on from generation to generation to make up the variety of humans and mutants on planet Earth.

Professor X's DNA could be thought of as a chemical ladder of just those four linked bases we mentioned above. The bases are strung together in pairs in a long and complicated sequence. As we said earlier, if you strung out the DNA of just *one* cell from Professor X's body, it would stretch, end to end, to about two meters long. But if you strung out the DNA from *every* cell in his body, end to end, it would reach from here to the Moon and back, three thousand times over. That's a lot of information.

So human beings are a kind of information recipe, passed on from generation to generation. The recipe is made up of billions of coded chemicals—the bases in DNA. When human bodies make new cells, they don't make many mistakes. But, once in a while, glitches occur. And sometimes, when the recipe is being copied, a single base pair gets added, left out, or exchanged for another base pair. These glitches are mutations, and we all have them. The X-Men have very notable mutations. And when mutations happen, they're passed on to the next generation, just as happened in the 2017 movie *Logan*, where an aged Wolverine defends his mutated Mexican daughter Laura, who has inherited powers.

This is how mutations occur: Imagine you're copying the longest book you can think of. And imagine you're copying it by hand. The book is so long, you're even working late at night. Even though you're very, very careful, paying lots of attention, and drinking lots of coffee, once in a while you make a spelling mistake. The same thing happens to Professor X's (and our) DNA, as it's passed on through the generations.

When we spoke earlier of evolution being a time bomb, we talked about how chemicals in the rocks can be used as markers to count the passing of historical time. Scholars can do the same with mutation markers. They help tell the tale of the Earth in human blood, and bone. Mutations are also a type of time machine. They can help us find out about our earliest ancestors. And they also help us expect the unexpected with future mutations.

THE DARWINIAN DOZEN: REAL-LIFE MUTANTS

So, Marvel apart, nature herself does a pretty good job at making mutants. After all, mutations are nature's way of creating new abilities and new traits. A good example is the chance mutation, probably around twelve thousand years back, which allowed some ancient X-Men to drink cow's milk and not barf. In time, this useful ability was passed onto many other humans. It helped save humans from starving, as long as you could catch a cow or two. Scholars think that every time the human genome is copied, there are around one hundred new mutations. The huge majority is, of course, rather dull and humdrum. But, every now and then, nature throws up some really remarkable glitches. In no particular order, here is a Darwinian Dozen, Charlie's X-Men—the real-life mutants whose glitches were useful.

CAPTAIN FINISH

Not quite Cap'n 'Murica, but more like Captain Finish. Eero Mäntyranta, the Finnish Olympic skiing champion, had genetic super endurance powers. Along with other members of his family, Mäntyranta "suffered" from a genetic condition, which resulted in a mutation of the erythropoietin receptor gene. In short, Eero was able to carry 50 percent more oxygen in his bloodstream, a trait that helped him become skiing champion. While not exactly Steve Rogers, such endurance always meant Mäntyranta was able to finish.

IN LIKE HULK

Nature plays a neutral hand, with glitches in the genome happening by pure chance. And that means that genetic "disorders" can turn out good, as well as bad. On the *good* side of that bell curve is Michigan lad, Liam Hoekstra. Liam has an extraordinary mutation that means his body doesn't make the protein myostatin, which inhibits muscle growth. Net result is that Liam's deficiency means he has much larger muscles, little body fat, and an inborn super strength. He's naturally stronger than his peers without the need for training, as long as he feeds his hungrier-than-normal body.

MATTER EATER MAN

In 1962, the DC Universe created a character known as Matter-Eater Lad. This superhero possessed the power to eat matter in all forms. Curiously, a dozen years before Matter-Eater Lad's first appearance in print, nature had produced its own version in Michel Lotito. Lotito was a French entertainer known as Monsieur Mangetout, which aptly translates to "Mister Eats All," though we might call him Matter Eater Man. Lotito suffered from Pica—a rare genetic condition that causes cravings and an appetite for things like dirt, glass and, I kid you not, anything metal. In normal humans, this mutation might lead to a blocked intestine, at the very least. But Lotito was "blessed" with a "very thick" lining in his stomach and intestines, which allowed sharp objects to pass through his system without doing any structural damage. Perhaps Lotito's most famous feat was consuming a whole Cessna 150 airplane, which he bolted down in bite-size chunks over a period of two years. Reading this is enough to give you indigestion.

SUPER EIDETIC MEMORY

Eidetic memory, sometimes called photographic memory, is the ability to acutely recall images from memory after just a few moments of exposure, without using a mnemonic device. In comic books such a skill was boasted by Batman and Bane, as well as Kal-El and Ozymandias, a.k.a. Adrian Veidt of the graphic novel, *Watchmen*.

But nature has conjured its own real-life eidetics. American actress Marilu Henner is a case in point. Marilu is perhaps best known for her role as Elaine in the popular sitcom, *Taxi*, though she also played Charles Boyle's love interest, Vivian Ludley, in another popular American sitcom, *Brooklyn Nine-Nine*. Marilu has hyperthymesia, which means she can recall particular details of practically every day of her life since she was small. And that includes precise details of filming *Taxi* for late-1970s TV. Marilu has described her memory ability to be similar to viewing "little videos moving simultaneously. . . . When somebody gives me a date or a year or something, I see all these little movie montages, basically on a time continuum, and I'm scrolling through them and flashing through them."

To date, scholars have identified just two dozen confirmed cases of hyperthymesia. Some scholars believe hyperthymesia is associated with an obsessive-compulsive need to constantly review, and renew, one's memories. But there's also a physiological link, in which the temporal lobe and caudate nucleus of the brain are enlarged in the eidetics.

FLEXIBLY FANTASTIC

Marvel's Reed Richards is a founding member of the Fantastic Four. Also known as Mister Fantastic, Richards's body mutated after being bombarded by cosmic radiation. The net result: Richards evolved the power to stretch his body into any shape he liked. A distant second to Reed Richards is the naturally flexible Spanish actor, Javier Botet. Fans of horror are more likely to have seen Javier at work than are fans of science fiction. Javier was the emaciated creature at the end of the superb Spanish horror film *REC 2*, as well as the titular character in the film *Mama* (though you might also catch him in the 2017 sci-fi movie, *Alien: Covenant*). You can mess with CGI as much as you like, but when you have a 6.6-foot tall actor, weighing only 123 pounds, and able to bend himself into the most ungodly poses, the natural spookiness of Javier Botet comes across far better than any old CGI camera tricks.

Botet's genetic condition is known as Marfan Syndrome. It affects the connective tissue of the entire body, rendering Marfan people with exceptional height, long limbs and fingers, and abnormal flexibility. Worryingly, Marfan is a spectrum disease, which means severe cases can be life-threatening to the heart and other organs. But those like Javier with milder symptoms live normal, healthy lives, scaring the living daylights out of cinemagoers.

SUPERHERO PARTYING ANIMALS

If you were to throw a superhero party, which of the usual suspects would be most likely to survive the ordeal? Tony Stark is perhaps the most famous alcoholic in comic book history. He'd also have the added benefit of being able to finance the party for you. Then there's Wolverine. Logan has been in so many bars in comic books over the years, it seems he's been in taverns more than he's been in trouble. Finally, there's the

Amazonian Wonder Woman. She's claimed her female tribe is well known as "Champion Drinkers," so they'd definitely make the cut.

Meanwhile, representing natural-born humans is Ozzy Osbourne. Yes, Ozzy is still alive. Even though he's partied to excess over many years, Ozzy has survived when many partying fellow rock musicians have perished long ago. Could Ozzy's secret be some kind of mutant partying superpowers, and a confidential status as an honorable underground X-Man? In 2010, Mr. Osbourne had his genome sequenced. Scholars revealed they'd found a number of gene variants never seen before. The variants are—surprise, surprise—present in those regions of Ozzy's genome associated with alcoholism and how the body absorbs recreational drugs.

A FACE THAT LAUNCHED A THOUSAND SHIPS

"Was this the face that launch'd a thousand ships/And burnt the topless towers of Ilium? Sweet Helen, make me immortal with a kiss," Christopher Marlowe wrote of Helen of Troy, or as Marlowe had it, "Helen of Greece." Thousands of years later and nature conjured up another beauty in British-American actress, Elizabeth Taylor. Ms. Taylor's undoubted loveliness was down, at least in part, to a mutation known as distichia. The rare condition meant that her eyelashes emanated from an unusual position on her eyelids. In essence, Ms. Taylor effectively boasted a double-set of eyelashes, which only helped folk focus on her famous "violet" eyes.

SLEEPLESS IN SAN FRANCISCO

Scientific wisdom has it that, on average, humans need about eight hours of sleep at night. But, some people, it seems, can get by on far less. According to scholars at the University of California, San Francisco, some sleep-test participants are "short sleepers." These mutants need a lot less sleep than the average Joe, due to a genetic disposition, which may affect around 5 percent of the population. Prime examples of this condition were a mother and daughter who shared an abnormal copy of a gene known as DEC2, which affects the circadian rhythm, the natural biological process that cycles every twenty-four hours.

MUTANT POLYDACTYLS

How's this for a genetic catch? Former baseball pitcher, Antonio Alfonseca, was renowned for his killer sinker. No doubt due in part to practice and perseverance, his skills must surely have also been owing a debt to an extra digit, for Alfonseca was born with a *sixth* finger on his pitching hand. This genetic mutation is called *polydactyly*, and means the suffering mutants are born with extra toes or fingers. Indeed, Alfonseca, a.k.a. *"El Pulpo"* or "The Octopus," has an additional digit on each hand and foot. That's quite modest when compared to young Indian lad, Akshat Saxena. Akshat was born with seven fingers on each hand and ten toes on each foot, which seems like a truly gripping genetic condition until you hear that the poor lad was born without thumbs.

Mutants like Alfonseca and young Akshat were revered in some ancient cultures. For example, the Pueblo people of Chaco Canyon decorated their great houses with six-digit footprints and sandal-shaped art. To find out how common mutant polydactyly was among the Chacos, scholars examined ninety-six skeletons excavated during expeditions. Incredibly, they found three polydactyl people among the ninety-six skeletons, and each polydactyl had an extra little toe on the right foot. This three out of ninety-six occurrences, which is around 3.1 percent of the Chaco population, is a much higher rate of polydactyly than in modern Native Americans, which stands at only 0.2 percent.

UNBREAKABLE BONES

The 2000 thriller movie, *Unbreakable*, directed by M. Night Shyamalan, was listed in *Time* in 2011 as one of the top ten superhero movies of all time. The film is about Elijah Price, born with Type I osteogenesis imperfecta, a rare mutation that means the sufferers' bones are very fragile and prone to fracture. In the movie, Elijah—who grows up to become a comic-book art dealer—develops a theory, based on comic books, that if he is at one extreme of human frailty, there must surely be other mutants who are "unbreakable" at the other extreme of the spectrum.

And indeed there are such real-life mutants. An unidentified family living in Connecticut was reported to have a genetically mutated condition, which meant members of the family had especially strong and dense bones. Not only that, but the bones seemed to be resistant to the wear and tear of time, like a kind of anti-osteoporosis.

PURE HEALING POWER

Toward the end of the 1970s, the virus later known as HIV affected the gay community. But one mutant managed to survive, even as friends around him fell to the then-mysterious disease. Scholars later found that Stephen Crohn had a "delta 32 mutation," which protected some of his white blood cells from HIV. This minutest of genetic changes meant that Crohn was totally immune. And this pattern is typical, for with any disease outbreak, a few mutated people in a population are usually found to have immunity for some reason. This happy fact of mutation often helps scholars define the disease and devise a cure.

LEAVE NO FINGERPRINTS

Some comic book fans have often wondered why Lex Luthor never demanded a comparison of the fingerprints of Clark Kent and the glove-shunning Superman. DC Comics editor Eddie Berganza provided an answer to the mystery when he suggested that Superman had perfected the ability to not leave any fingerprints behind.

Such a genetic condition actually does exist. Known as "immigration delay disease," adermatoglyphia is a mutation that renders the afflicted fingerprint-less. The condition was first uncovered when a Swiss woman was refused entry to the United States because of her mutation—all non-residents are required by law to be fingerprinted. Apart from the fact that this condition would result in rather Kafkaesque dealings with various bureaucracies, this rare mutation might also gift the sufferer an essential power to become a super-criminal!

PART III
MACHINE

HOW MIGHT A PREPPER MAKE AN IRON MAN SUIT?

"I do not remember how it got into my head to make the first calculations related to rocket. It seems to me the first seeds were planted by famous fantaseour, J. Verne."

—Konstantin Tsiolkovsky, quoted in *LIFE: 100 People Who Changed the World* (2010)

"It was the first time I'd had to design something that saved lives. It was a stopgap at best. I got home and put my money into a suit that'd keep me alive . . . But I kept the suit. Kept tinkering with it. And I'm not sure why any more. Except maybe that it wasn't about the future. But my future. And it allowed me to pretend that I wasn't a man who made landmines. I went from being a man trapped in an iron suit to being a man freed by it. [Iron Man command system on. Start.]"

—Tony Stark in *Iron Man: Extremis,* written by Warren Ellis (2005)

"Hey," Watney said over the radio, "I've got an idea."

"Of course you do," Lewis said. "What do you got?"

"I could find something sharp in here and poke a hole in the glove of my EVA suit. I could use the escaping air as a thruster and fly my way to you. The source of thrust would be on my arm, so I'd be able to direct it pretty easily."

"How does he come up with this shit?" Martinez interjected.

"Hmm," Lewis said. "Could you get 42 meters per second that way?"

"No idea," Watney said.

"I can't see you having any control if you did that," Lewis said. "You'd be eyeballing the intercept and using a thrust vector you can barely control."

"I admit it's fatally dangerous," Watney said. "But consider this: I'd get to fly around like Iron Man."

"We'll keep working on ideas," Lewis said.

"Iron Man, commander. Iron Man."
—Andy Weir, *The Martian* (2011)

"... with the equally successful Avengers series. That clown doesn't have half your talent . . . and he's making a fortune in that Tin Man get-up."
—Riggan Thomson, *Birdman: Or (The Unexpected Virtue of Ignorance)* (2014)

You've heard of Pepper Potts. Meet "Prepper" Potts. Prepper's doomsday dreams come true. The world had come to an end. Well, most of it anyhow. He was alone, as usual, practicing his prepping protocols in the mountains, when the calamity struck. An especially infectious rogue pathogen managed to breach the barrier between species and set up home in human hosts. Or maybe it was a conscious act of bioterrorism. You never can tell with corporations these days; they don't even pay their taxes, but are more than happy to fund the world's wars.

Whatever the origin, the contagion spread like wildfire. In these days of cities packed as dense as nuclear stars, and intercontinental air travel as common as comets, the world's population was picked off as easily as a conjurer swallowing a poker. Most were dead before Big Pharma could develop its vaccine and quarantine, and delightedly count their dollars as if they fell from heaven itself. Now, even Big Pharma is dead.

Prepper returned from the remote hills after a self-imposed exile and found to his delight that civilization no longer existed. Cities, highways, and homes were deserted. It was a dream come true. Most "normies" would struggle with coming to terms with their new predicament—how could they possibly cope with the collapse of the whole infrastructure that used to shore up civilization? What can they do in this hopeless situation to ensure their own survival?

And yet none of this concerns Prepper. He's fit for the superhuman task of rebuilding the world from scratch. He's survived the apocalypse. Now he has to survive the aftermath. Prepper has read all the survivalist handbooks. But he's not yet concerned with the rebuilding of civilization, cultivating crops, or making clothes. Before all of that he has to survive the roving gangs of scavengers, and the psychopaths who prey ruthlessly on the less well organized, and the unarmed. A civilized person might wonder if morality and sanity would prevail. But it seems this brave new world is full of preppers who've watched *Mad Max: Fury Road* on repeat, and naturally taken its crazed philosophy entirely to heart.

What Prepper really needs is an Iron Man suit. In this wasteland where marauders roam, only the maddest will prevail. Or, failing madness, someone kitted out like a post-apocalyptic Tony Stark. But Prepper simply doesn't have the luxury of being too ambitious. If he's going to rebuild the world from scratch, then the same applies to the suit, and the best he can hope for at first is Iron Man Armor Mark I. After all, like Prepper, Tony Stark built his first suit from the limited resources he found around him in that cave in Afghanistan. He salvaged scraps of metal and machine parts and synthesized them into something special. Like Stark, Prepper hopes that a suit will enhance his strength and durability. It'll help him overpower armed marauders. If fashioned well, the many layers of durable metal will make his armor impervious to the kind of medium-caliber firearm attack you could expect out there in the wasteland.

PREPPER POTTS TAKES TO THE AIR

But how would his suit handle the kind of flight managed by Iron Man? Flight would be a great advantage for survival. Prepper checks his survivalist manuals and learns that something similar to an Iron Man jetpack can be built from using downward-firing machine guns. The idea seems simple enough. Since a bullet fired forward kicks back with recoil, a gun fired downward must surely push you up. Guns may be able to lift their own weight, when fired. A machine gun that weighs only ten pounds, but produces a recoil equivalent to twelve pounds, is going to lift itself slightly off the ground.

So Prepper gets into the dynamics of what engineers call, thrust-to-weight ratios. A ratio less than one means that a craft's thrust won't be enough to lift it off the floor. But Prepper checks out the ratios of prime and past examples: The Rolls-Royce/Snecma Olympus 593 turbojet that powered the supersonic airliner Concorde had a thrust-to-weight ratio of 5.2. And the Merlin 1D engine developed by SpaceX back in 2012 had a thrust-to-weight ratio of 180.1! Far too ambitious for doomsday, clearly. But he discovers something useful about the AK-47, also known as the Kalashnikov. After almost seven decades, the AK-47 and its variants remain the most popular and widely used assault rifles in the world. He may well find plenty AKs out in the wasteland. Not only that, but the AK has a thrust-to-weight ratio of two. If Prepper stood an AK on end and managed to tape down the trigger, the AK might take off like a Soyuz while firing.

In practice, Prepper finds the AK performs even better. And he works out the reason why. The amount of thrust made by a rocket or a machine gun usually depends on (a) how much mass it's spitting out, and (b) how fast it's spitting. But when Prepper does his AK tests, he finds the actual thrust is roughly a third higher again because the gun is spitting out fiery debris and gas, as well as the expected bullets. And so, Prepper figures, the AK-47 might well take off, but its spare thrust would be too little to lift anything more than a kitten. Hardly the manliest of conclusions, so back to the drawing board.

What about a jetpack that uses multiple machine guns? If he finds enough AKs out in the field he might be able to strap them together. Two AKs fired down at once would make twice the thrust, so if each gun could lift 3 pounds more than its own weight, say, then two guns would lift 6 (Prepper didn't neglect basic math in preparing for doomsday). But how many AK-47s is Prepper going to have to add before he can Tony Stark it across the deserted plains of this brave new world? At this rate, he's going to end up with something that looks more like a hovercraft than a jetpack.

Prepper's next problem is the question of ammunition. Okay, granted, he still needs to find enough AK-47s out in the wasteland, manage to tape the triggers down, when needed, and somehow construct some kind of raft of guns to sit on. But an average AK magazine houses only thirty rounds. And if the gun is blasting out at ten rounds each second, that only gives Prepper three seconds of acceleration. He finds a partial solution when he learns that larger magazines can carry more ammo. But then the rocket science books suggest that carrying any more than around 250 rounds would be a disadvantage, as all that "fuel" would simply make you too heavy to take off.

After applying a little more math, Prepper comes to a conclusion about his best-fit AK craft. Ideally the craft would feature about three hundred AK-47s (the problem remains of sourcing that kind of fire power!), each carrying 250 rounds of ammo. Prepper calculates such an AK craft would zoom him up to speeds of around 200 miles per hour, and an altitude of a third of a mile. Enough to escape the marauders, but such an impractical craft that Prepper risks making an appearance in some kind of post-apocalyptic Darwin Awards, that tongue-in-cheek honor that recognizes those who contribute to human evolution by selecting themselves out of the gene pool, via death or sterilization by their own idiotic actions.

(Incidentally, Prepper's favorite Darwin Award story was that of Larry Walters. Larry lived in LA. His boyhood dream was to fly, but poor eyesight disqualified him from the job of pilot. To make amends at this life's failure, Larry bought forty-five weather balloons from an Army-Navy surplus store. Then, he tied them to his lawn-chair and filled the four-foot diameter balloons with helium. Armed with some sandwiches, Miller Lite, and a pellet gun, Larry strapped into his lawn-chair, and took flight. He calculated to shoot a few balloons when it was time to descend. But Larry's plan to lazily float to a height of about 30 feet above the backyard went wrong. He rocketed into the LA sky, and finally leveled off at a mere 16,000 feet. Up in the gods, Larry simply couldn't risk shooting any of the balloons, just in case he unbalanced the load and found himself in real plummeting trouble. So he drifted, frozen and terrified along with his beer and sandwiches, for more than fourteen hours. Eventually, Larry crossed the primary approach corridor of LAX, where startled pilots radioed in reports of the strange and surreal sight of Larry and his wacky balloon ship). Prepper simply couldn't risk something similar with his AK craft. In doomsday, there just weren't any pilots around to save your ass. So Prepper drops the AK-47 and considers heavier firepower.

THE AVENGER AUTO-CANNON JETPACK

An auto-cannon might do the trick. General Electric had been making a Gatling-type auto-cannon since 1977. It was known as the GAU-8 Avenger, and was used before "the fall" on the Fairchild Republic A-10 Thunderbolt

II jets, as well as the Goalkeeper CIWS ship weapon systems. And what got Prepper really excited was the Avenger's clout. The Avenger fired as many as sixty one-pound bullets, each and every second. Even more remarkable, this firepower meant that the Avenger packed almost 5 tons of recoil force. This was even more incredible when Prepper discovered that the A-10 jets that used the Avenger had two engines that each produced only 4 tons of thrust. That meant if you mounted two Avengers on one A-10 jet, and blasted both guns forward at the same time as engaging the jet's throttle, the Avengers would triumph and the A-10 would move backward.

All this was perfect news for Prepper. He momentarily forgot about his Iron Man suit and daydreamed about mounting an Avenger on top of his Land Rover. He worked out that, mounted on his car firing backward and with the Land Rover in neutral, the firepower of the Avenger would mean his old ride would go from rest to breaking the speed limit in less than three seconds. This prospect reminded Prepper of another, apocryphal, Darwin award winner. A former Air Force sergeant had fixed some JATO (Jet Assisted Take-Off) solid fuel rocket units to his 1967 Chevy Impala and found a long, straight stretch of road in the Arizona desert. When the JATOs fired, the Impala swiftly reached a speed of 300 mph and the car became airborne for over a mile, before impacting a cliff face at a height of 125 feet, leaving a blackened crater three feet deep in the rock. Arizona Highway Patrol said the metal debris at the scene resembled the site of an airplane crash. No sign of the Impala owner could be found.

And yet he had his solution. If he could source an Avenger, make his suit sturdy enough to survive the blast power, and wrap the Avenger in some kind of aerodynamic skin, Prepper could make like Iron Man.

HOW DOES THOR'S HAMMER WORK?

"Whosoever holds this hammer, if he be worthy, shall possess the power of Thor."

—Ashley Edward Miller, Zack Stentz, and Don Payne, *Thor* screenplay (2011)

Interior. Party scene. Stark tower.

After enjoying the evening, the Avengers relax as Thor places his hammer, Mjölnir, on the coffee table. A challenge of sorts is issued.

TONY STARK: "Never one to shrink from an honest challenge . . . It's physics."

BRUCE BANNER: "PHYSICS!"

[Stark grasps Thor's hammer]

TONY STARK: "Right, so, if I lift it, I . . . I then rule Asgard?"

THOR: "Yes, of course."

TONY STARK: "I will be re-instituting prima nocta."

[Stark tries to lift the hammer but fails]

TONY STARK: "I'll be right back."

[Wearing his armored hand, Stark tries to lift the hammer again and fails. Then, wearing their armored hands, Stark and James Rhodes both try to lift Thor's hammer]

JAMES RHODES: "Are you even pulling?"

TONY STARK: "Are you on my team?"

JAMES RHODES: "Just represent! Pull!"

TONY STARK: "Alright, let's go!"

[They both pull as hard as they can. Next, Banner tries to lift the hammer, he roars trying to change to the Hulk but fails, and everyone but Natasha stares at him warily. Natasha grins.]

BRUCE BANNER: "Huh?"

[Next, Steve Rogers gets up to try]

TONY STARK: "Let's go, Steve, no pressure."

JAMES RHODES: "Come on, Cap."

[Steve starts pulling on the hammer and manages to budge it a little; Thor looks a little alarmed. Steve still fails to lift it; Thor laughs with relief].

—Joss Whedon, *Avengers:Age of Ultron* screenplay (2015)

"Galileo's dialogue concerning the two chief world systems is a discourse between three characters, carried out over four days, and written in Italian. From the outset of the dialogue, the very naming of its three main characters speaks volumes of the stance that Galileo was to adopt in the rest of the book. Salviati, the brilliant academician, is Galileo's advocate. Sagredo, the intelligent layman, is, at least initially, neutral. And Simplicio, dedicated follower of Aristotle, Ptolemy, and therefore the Church, is the ignorante, the well-meaning muggins, who understands little of the world in which he lives. It is hardly surprising that the Inquisition would object to this last name's resemblance to 'simpleton.'"

—Mark Brake, *Revolution in Science: How Galileo and Darwin Changed Our World* (2009)

AVENGERS: AGE OF ULTRON (MY TAKE ON A DELETED SCENE)

All comers have been defeated in the challenge to lift Thor's hammer, *Mjölnir*, off the coffee table. Three remaining Avengers, Stark, Banner, and Rogers, talk in Thor's absence. The topic—how might the hammer actually work, and how might the Mjolnir challenge be met and overcome. Earlier that day, Stark and Banner had been chatting about Galileo's way of putting ideas over in print. He'd often used the form of a dialogue between protagonists, in order to try working out the best way of thinking about a certain situation. As the conversation unfolds, it is unclear which one of Stark or Banner is Salviati, and yet neither care to admit that both are left with the rather uncharitable thought that Rogers is certainly the person most likely to play the role of the simpleton. The discussion begins . . .

TONY STARK: "So what's the secret to beating Thor's hammer challenge?"

BRUCE BANNER: "It's got to be gravity and mass."

TONY STARK: "Let's back up a little there, Bruce. Before we get into the gravity of the situation, what's this thing actually made of?"

STEVE ROGERS: "Well, guys, Thor told me that Mjölnir is forged by Dwarven blacksmiths, and is made of some Asgardian metal called uru. The dwarves must have also forged the hammer's inscription of "Whosoever holds this hammer, yada yada yada."

TONY STARK: "I don't buy it."

STEVE ROGERS: "You don't buy yada yada yada?"

TONY STARK: "No, I don't buy it, the uru thing. How can there be a naturally occurring metal on Asgard that isn't in the rest of the universe?"

STEVE ROGERS: "Magic?"

TONY STARK: "Go to the bottom of the class, Cap."

STEVE ROGERS: "Funny. Wait a minute, Thor ALSO told me that Mjölnir was forged in the heart of a dying star."

BRUCE BANNER: "*In* the heart of a dying star, or *from* the heart of a dying star?"

STEVE ROGERS: "Is there a difference?"

BRUCE BANNER: "Well that's a pretty roasting environment, Steve, the heart of a star. Even our Sun's core temperature sits at about fifteen million degrees. Those dwarves would get singed some."

TONY STARK: "But *from* the heart of a dying star, or from dead star material, that's another matter—ha!"

STEVE ROGERS: "How so?"

BRUCE BANNER: "Some stars die down, collapse under their gravity, and shrink to only a few miles across."

TONY STARK: "Neutron stars."

BRUCE BANNER: "Right. And these stars, kinda dead, are made of neutrons and nothing else."

STEVE ROGERS: "They're dense?"

BRUCE BANNER: "And then some. They're the smallest and densest stars known to exist."

TONY STARK: "Neutron stars are so dense that one teaspoon of their stuff would have a mass around 900 times the mass of the Great Pyramid of Giza."

STEVE ROGERS: "That'll do it!"

TONY STARK: "It explains how Thor can plant Mjölnir on the lower jaw of a dragon and pin it to the ground, shouting "Stay!" though I ain't seen that for myself."

STEVE ROGERS: "Be patient. But how does this help us with the challenge?"

BRUCE BANNER: "Time to talk about gravity and mass."

STEVE ROGERS: "What about gravity?"

TONY STARK: "It helps explain how things fall in gravity fields, including hammers."

STEVE ROGERS: "Zak Newton and those guys?"

TONY STARK: "Sure. Potted history: Aristotle, ancient Greek guy, beard, Jesus sandals, figures that falling objects are just trying to find their natural resting place. And for hammers, that would be the center of the Earth, or planet."

BRUCE BANNER: "Yeah, and Isaac Newton thought that all masses, including hammers, create a gravity field in the space around them."

TONY STARK: "Zak was slightly shady, though, and borrowed some of his ideas from the occult, which brings us to Einstein and his idea of gravity as a warp in space-time. So, Thor's hammer could possibly emit gravitons. Okay, now gravitons are hypothetical particles that mediate the gravity field. If Mjölnir could recognize Thor, through some kind of nanotech, it could speedily change its mass. [TONY clicks his fingers] [CLICK] Now Mjölnir is light enough for Thor to pick it up, but [CLICK] now it's too heavy even for Bruce in Hulk mode to lift it, and when Bruce is Hulk, he can run through an aircraft carrier."

STEVE ROGERS: "Yeah? But there have been rumors in the past that others have managed the challenge."

TONY STARK: "Nanotech glitch?"

BRUCE BANNER: "What about the Higgs Boson?"

STEVE ROGERS: "The what now?!"

TONY STARK: "The so-called God particle."

BRUCE BANNER: "The Higgs Boson is meant to be the particle that provides mass to the most basic building blocks of matter. It's key to the fundamental forces and particles of the universe."

STEVE ROGERS: "But how does this Higgs Boson help with the hammer challenge?"

TONY STARK: "The Higgs is meant to be the particle that gives mass to matter."

BRUCE BANNER: "Well, it doesn't technically give other particles mass. The Higgs Boson is a quantized manifestation of the Higgs field, which generates mass through its interaction with other particles."

STEVE ROGERS: "Is this really getting us anywhere at all, guys?"

BRUCE BANNER: "Sure! Imagine Mjölnir has a signature button for Thor's thumbprint, like the Touch ID on the home button of an iPhone."

STEVE ROGERS: "Mjölnir has Touch ID!"

BRUCE BANNER: "Well, could be. Thor's thumbprint would switch the Higgs field off for Thor . . ."

TONY STARK: "So the hammer becomes feather-light . . ."

BRUCE BANNER: ". . . right, and when Thor's signature isn't present, the hammer reverts to its usual 900 times Great Pyramid of Giza, heart of a dying star, dragon-jaw dropping mass."

TONY STARK: "Anyone know any good Touch ID hacks?"

WHY MIGHT PLAYING SUPERHERO VIDEO GAMES HELP SAVE THE WORLD?

"Reality is broken. Game designers can fix it."

—Jane McGonigal, *Reality is Broken: Why Games Make Us Better and How They Can Change the World* (2012)

"Game developers know better than anyone else how to inspire extreme effort and reward hard work. They know how to facilitate cooperation and collaboration at previously unimaginable scales. And they are continuously innovating new ways to motivate players to stick with harder challenges, for longer, and in much bigger groups. These crucial twenty-first-century skills can help all of us find new ways to make a deep and lasting impact on the world around us."

—Jane McGonigal, *Reality is Broken: Why Games Make Us Better and How They Can Change the World* (2012)

"I want gaming to be something that everybody does, because they understand that games can be a real solution to problems and a real source of happiness. I want games to be something everybody learns how to design and develop, because they understand that games are a real platform for change and getting things done. And I want families, schools, companies, industries, cities, countries, and the whole world to come together to play them, because we're finally making games that tackle real dilemmas and improve real lives."

—Jane McGonigal, *Reality is Broken: Why Games Make Us Better and How They Can Change the World* (2012)

"The scene opens up to a dark, stormy night. Dark clouds pass, and the Bat Signal is seen illuminated. A statue of the Grim Reaper slowly rises as the view changes to a busy street. A radio transmission from a nearby police car is heard.

Radio: All units to Gotham City Hall. The Joker has been apprehended. Batman is now en route to Arkham Island . . ."

—Introduction to *Batman: Arkham Asylum* video game (2009)

Fancy being Batman? You can inhabit the very being of the Dark Knight himself. You can go, with inhuman stealth, through the grim, gothic streets of Gotham, down to Arkham Island, home of the asylum. You can witness at first hand the mental autopsy on Batman and his enemies, experience the journey of Bruce Wayne on the road to greatness. In the face of your foes, you will be severely tested. Your arch-nemeses, adversaries such as the Joker and Two-Face, will tear your mental straightjacket apart. And, before the battle is done, your every weakness will be exposed on the eve of your eventual triumph. Or maybe you fancy the more vertiginous thrills of being Spiderman. If so, scaling walls and swinging freely between skyscrapers is totally possible in video games. In a bright isometric and virtual world, you vie for supremacy with a huge roster of infamous friends and foes, from Doc Oc and Daredevil to Venom and Rhino. And soon you'll have the chance of playing as one of the Avengers. In 2017, the games world announced it had signed a multi-game partnership with Marvel. The plan is to produce a stunning series of Avengers titles, steered by two reputable studios: Crystal Dynamics, responsible for titles such as *Tomb Raider*, and Eidos Montreal, creators of the *Deus Ex* series of games. So the prospects are excellent, playing superheroes in big, open world action adventures, with a good sprinkling of role-playing gravitas.

But is gaming doing you any good? You may be familiar with the kind of propaganda that spouts out from conservative print and electronic media. "Games are melting your brain"; "games are ruining your eyes"; "games are turning you into a violent person," etc. (Clearly, society outside of the virtual world isn't violent in the slightest.) When desperate to prove their point, they even attempt to invoke some actual science to back up their bias. A study that purports to show that playing "shooter" games may damage the hippocampus area of the brain, causing it to hemorrhage cells; or the untestable contention that gaming could weaken the brains of the young, putting them at greater risk of dementia in later life; or the countless studies that in truth offer very conflicting conclusions on how detrimental video gaming is to health—both mental and physical.

It all amounts to a kind of "games curse." For millennia, they said something similar about books. The use of book curses dates back to pre-Christian times, when the wrath of the very gods themselves was invoked to protect books and scrolls. "He who breaks this tablet or puts it in water or rubs it until you cannot recognize it [and] cannot make it to be understood, may . . . the gods of heaven and earth and the gods of Assyria curse him with a curse that cannot be relieved, terrible and merciless, as long as he lives, may they let his name, his seed, be carried off from the land, may they put his flesh in a dog's mouth," said one such book curse from Babylonian times. Warnings about books became even more widely employed after the invention of the printing press. Medieval conservatives warned of the isolationist and antisocial nature of taking oneself off into a corner and enjoying an educating read. Such suspicions about the evil of books was still in the culture centuries later, 1856 to be precise, when Gustave Flaubert wrote *Madame Bovary.* "Reading novels and all kinds of bad books . . . It's a dangerous business, son . . . So, it was decided to prevent Emma from reading novels. The project presented certain difficulties, but the old lady undertook to carry it out: on her way through Rouen she would personally call on the proprietor of the lending library and tell him that Emma was cancelling her subscription. If he nevertheless persisted in spreading his poison, they would certainly have the right to report him to the police."

At one time, even chess was laughably thought to encourage violence. Although chess has been enjoyed innocently for centuries, and is not normally associated with swordplay or gun-toting gangsters (unless I'm missing an important memo), it wasn't until the nineteenth century that the game truly expanded into Europe and the US. Conservatives were, naturally, suspicious of this "crazy" new fad, and several critics denounced the game as a source of intellectual laziness and anti-social behavior, which could even provoke violence.

VIDEO GAMES BITE BACK

This all obscures the fact that video games, like reading and chess, have untold value. And that value is increasing. Sure, gaming has always been popular. And this popularity has truly soared over the last few years. Like the comic books and science fiction sub-genres that underpin many titles, games were once branded with clichés of being seen as "geeky" and "uncool." But nowadays, geeky *is* cool. Of the top twenty highest-grossing films of all time, over half are science fiction. And there's the huge rise of comics and "cosplay" Comic-Con conventions, which has helped contribute to a massive change in perception. When almost everyone is a geek, clichéd labels no longer work.

Within this changing culture, gaming has become increasingly dominant. Incredibly, in 2017, the amount spent on video games is projected to reach $92 billion. That's more than we spend on movies ($62 billion) and recorded music ($18 billion) combined. As interactive games have got more massive, the more passive experience of music and movies seems to have been left behind. The contrast is poised to become even starker, as the rise of virtual reality games becomes more dominant. And why has the games industry become five times larger than the music industry, and around one and a half times larger than the movie industry? Consider gaming's more recent advances.

Game culture has made seven great leaps in the twenty-first century. And this period of extraordinary change has helped facilitate cooperation and collaboration to previously unimaginable scales. Let's look at those seven advances and consider the way in which the games case has been strengthened.

The first great leap is the rise of open-world game design. This advance has seen the entire philosophy of game design evolve from the highly inflexible linear experience of game levels to a more playful experimental space. The new spatial composition of open-world games has inspired players to be more creative and curious, more audacious and collaborative. The second great leap is the rise of the independent games community. With the facility of the likes of Xbox Live, the App Store, and Steam, a worldwide audience for independent titles has been created. Simultaneously, academics are getting in on the act. The scholarly study of games has evolved, such as the new courses at the Department of Game Design at New York University's world-famous Tisch School of the Arts, and the UK's National Film and Television School's masters in game design. These developments have meant the creation of independent games that are idiosyncratic and eccentric, from the wholly subjective to the profoundly political.

Another important leap has been the rise of gaming as a social experience, with shared social spaces from which the community has benefited hugely. These experiences have included asynchronous competition, multiplayer games that can be played at your convenience, and anonymous collaboration. Added to that is the advent of friends lists, matchmaking, and huge online get-togethers, enabling gamers to meet up and play, or simply to chat. These first three leaps have led to another advance, namely the growing influence of games on the wider culture. Video games have now made appearances at The Museum of Modern Art in Midtown Manhattan, and at London's Tate and Victoria and Albert Museums. Television and movie productions derive much from the aesthetics and organization of games. And one of the best recent TV sci-fi series, *Westworld*, is a filmic discourse on the morals of virtual violence and player accountability.

The growing influence of games on the wider culture has meant that video games are being created and played by a greater and broader variety of people than ever, our fifth leap. Millions of commuters travel to work engrossed in handheld adventures, while more intimate communities explore themes such as identity, gender, and sexuality issues through visual novels, with identifiable heroes, like you, now part and parcel of the game narrative. Throughout history, from Greek tragedy to modern cinema, the "stage," as it were, has been a mirror, as well as a window. And this vital element of representation is happening in games.

Our sixth leap in gaming culture is player creativity and collaboration. Time was that, if you wished to make your own game levels, you had to have considerable technical savvy, and be truly obsessed. But games like *LittleBigPlanet*, *Minecraft*, and *The Sims* changed all that. Now, user creativity rose to the surface, as huge online communities grew up around building and sharing game content. No more games that are quickly consumed and soon discarded. Many titles have become global creative workshops. And emergent broadcast platforms, such as *YouTube* and *Twitch*, have triggered a fresh form of entertainment—the "let's play" video—to prosper, providing new ways for gamers to share their experiences.

And that brings us to our seventh and final leap in gaming culture: the exploration of new themes. Along with the evolution of gaming, the gamers have themselves matured. Game designers are getting wiser as they age, far more thoughtful, and their new titles are a cultural echo of this evolving maturity. Not only that, but a new generation of gamers are entering the industry, a generation that innately knows games as a form of cultural self-expression, rather than mere entertainment and commodity fetishism. In the course of this new century of ours, games have shifted from the chauvinist trope of saving the maiden from the dark tower, and towards more sophisticated notions of salvation, responsibility, and self-discovery. These days you can visit Steam and get downloadable content of a superhero sci-fi shoot-'em-up, or else consider time and causality in *Life is Strange*, a game in which an eighteen-year-old photography student discovers she has the ability to rewind time at any moment, which leads her every choice to enact the butterfly effect.

REALITY IS BROKEN

So the twenty-first century has become a world full of games. Hundreds of millions of people across the globe are gamers. And the average young person will spend over 10,000 hours gaming by the age of twenty-one. This state-of-the-world report has led some to believe the future belongs to the gamers. They believe that this groundbreaking century of extraordinary creativity and change in video games, where the rules, structure, and mechanics of games have been constantly ripped up and re-written, has the potential not only to radically improve our lives, but also to change the world.

One such believer is Jane McGonigal, visionary game designer and the Director of Game Research and Development at the Institute for the Future, in Palo Alto, California, a not-for-profit think tank, which was established in 1968 as a spin-off from the RAND Corporation. In her trail-blazing 2012 book, *Reality is Broken: Why Games Make Us Better and How They Can Change the World*, McGonigal challenges conservative thinking and shows that games—far from being mere escapist entertainment—have the potential of being parlayed into a recipe for changing the offline, "real" world.

Brilliantly deconstructing good game design and outlining the likes of the seven leaps in game advance we detailed above, McGonigal suggests that game designers stand ahead of the curve on inspiring extreme effort, and rewarding hard work. Through the leaps in game evolution, people's collaboration and cooperation have been orchestrated to previously unimaginable scales. And, crucially, continued innovation is finding new ways to work on harder challenges, and in much larger groups.

McGonigal believes these critical twenty-first-century skills will help forge new ways of making profound and lasting influence on the world around us. She envisages gaming as something that all of us will learn to do. From close-knit communities to a world-wide network of gamers. And this evolution will mean games become a real resolution to some of the world's problems, as well as a source of pleasure. *Reality is Broken* concludes, "We can no longer afford to view games as separate from our real lives and our real work. It is not only a waste of the potential of games to do really good—it is simply untrue. Games don't distract us from our real lives. They fill our real lives with positive emotions, positive activity, positive experiences, and positive strengths. Games aren't leading us to the downfall of civilization. They're leading us to its reinvention. The great challenge for us today, and for the remainder of the century, is to integrate games more closely into our everyday lives, and to embrace them as a platform for collaborating on our most important planetary efforts." Game over? Far from it.

THE PHYSICS AND FUN OF BEING DOCTOR MANHATTAN

"There is no science in this world like physics. Nothing comes close to the precision with which physics enables you to understand the world around you. It's the laws of physics that allow us to say exactly what time the sun is going to rise. What time the eclipse is going to begin. What time the eclipse is going to end."—Neil deGrasse Tyson

"The greatest progress is in the sciences that study the simplest systems. So take, say, physics—greatest progress there. But one of the reasons is that the physicists have an advantage that no other branch of sciences has. If something gets too complicated, they hand it to someone else."—Noam Chomsky

"Doctor Manhattan, as you know the Doomsday Clock is a symbolic clock face analogizing humankind's proximity to extinction, midnight representing the threat of nuclear war. As of now it stands at four minutes to midnight. Would you agree that we are that close to annihilation?"—Janet Black, to Doctor Manhattan, Watchmen movie (2009)

MANHATTAN: WHAT'S IN A NAME?

When it comes to being named Doctor Manhattan, quite a lot. Dr. Jonathan Osterman was a nuclear physicist who was transformed in a fictional 1959 into arguably DC Comics' most supreme being. Following his molecular disintegration and reconstruction, Osterman was employed by the United States government, who christened him Doctor Manhattan, after the Manhattan Project. The name was very apt.

For the Manhattan Project was the result of a science fiction story, written by H. G. Wells. The mysteries of the atom had long been a scientific Holy Grail. By the first light of the twentieth century it was clear that some form of atomic energy must be responsible for the tremendous power at the heart of the Sun and stars. In 1903, nuclear physicist Ernest Rutherford and his co-worker, Frederick Soddy, had been the first to calculate the vast amount of energy released in radioactive decay. Both were awake to the idea that this energy was potentially lethal. Indeed, as Rutherford said, "Some fool in a laboratory might blow up the universe unawares." And Soddy, in a lecture the following year, reflected that, "The man who put his hand on the lever by which a parsimonious Nature regulates so jealously the output of this store of energy would possess a weapon by which he could destroy the Earth if he chose." Soddy trusted nature to "guard her secret." H. G. Wells begged to differ.

Wells's novel *The World Set Free* (1914) led non-stop to the launch of the Manhattan Project. Wells's book actually coined the phrase "atomic bomb": ". . . And these atomic bombs which science burst upon the world that night were strange even to the men who used them." Medieval legends had professed that those of tainted spirit drinking from the Holy Grail would face instant annihilation. Wells was aware that the Grail of the atom offered the opportunity for great good or sheer evil.

On the eve of the First World War, Wells presented an ill-omened vision of future warfare. The book foresaw a holocaust where the world's key cities are annihilated by small atomic bombs dispatched from airplanes. This is no mere guesswork on the part of Wells. The weapons portrayed are truly nuclear; Einstein's equivalence of matter converted into fiery and explosive energy triggered by a chain reaction.

There had been earlier fiction on super-weapons. They had fallen prey to the cliché of comic books and pulp fiction—the naïve notion that the tangential mind of a single genius could change the course of history. Human problems could be solved by the techno-fix of a scientific miracle. Wells was wise enough to realize that the level of technical advance doesn't come from the know-nothing notion of genius. It comes from the dialectic between nations and their productive forces. Wells here predicted the emergence of the military-industrial complex.

His schedule for the development of nuclear capability is astoundingly accurate. In *The World Set Free*, the 1950s scientist who uncovers atomic energy realizes there is no going back. Nonetheless, he feels "like an imbecile who has presented a box of loaded revolvers to a crèche". Initially, nuclear capability merely leads to a greater strain on the system. The rich get richer, unemployment soars, crime rockets.

Global tensions become menacing, with governments "spending every year vaster and vaster amounts of power and energy upon military preparations, and continually expanding the debt of industry to capital." These contradictions of capitalism, which Wells brands a barbaric form of society, lead to nuclear holocaust. The Earth is scorched. The swarms of survivors, many mutilated by fallout and radioactive dust, wander the barren landscape in scenes now familiar in film and fiction.

Science still emerges as, "the new king of the world," however. In Wellsian fashion, a republic of mankind, governed by intellectuals and scientists, is established from the ruins of capitalism. Looking backward from this post-apocalyptic utopia to the mid-twentieth century, Wells delivers a damning indictment of his own time. "They did not see it until the atomic bombs burst in their fumbling hands. Yet the broad facts must have glared upon any intelligent mind. All through the nineteenth and twentieth centuries the amount of energy that men were able to command was continually increasing. Applied to warfare that meant that the power to inflict a blow, the power to destroy, was continually increasing."

"I AM BECOME DEATH. THE DESTROYER OF WORLDS"

Wells's fictional Bomb led straight to the Manhattan Project and, ultimately, Hiroshima. His visionary novel was the guiding inspiration for the brilliant Hungarian physicist Leo Szilárd. After reading *The World Set Free* in 1932, Szilárd became the first scientist to seriously examine the science behind the creation of nuclear weapons. "The book made a very great impression on me," Szilárd recalled. Thirty years later he still remembered the prophetic book. ". . . a world war . . . fought by an alliance of England, France, and . . . America, against Germany and Austria, the powers located in the central part of Europe. [Wells] places this war in the year 1956, and in this war the major cities of the world are all destroyed by atomic bombs."

Szilárd was a survivor of a devastated Hungary after the Great War. He had developed a lasting humanitarian passion for the protection of life and freedom, particularly the freedom to communicate ideas. Wells's book echoed in Szilárd many of the utopian beliefs that guided him in the years to come. A quiet determination now changed the direction of his work. Szilárd went into nuclear physics because he wished to contribute something to save mankind, "only through the liberation of atomic energy could we obtain the means which would enable man not only to leave the Earth, but to leave the solar system."

Szilárd became the driving force behind the Manhattan Project. A year after reading Wells's book, Szilárd fled to London to escape Nazi persecution. There he read an article in *The Times* by Rutherford. The professed "father" of nuclear physics, and pioneer of the orbital theory of the atom, Rutherford rejected the idea of using atomic energy for practical purposes. A legendary quick thinker, Szilárd was so incensed at Rutherford's dismissal that he dreamt up the idea of the nuclear chain reaction while waiting for traffic lights to change on Southampton Row in Bloomsbury, London. One year later he filed for a patent on the concept. It was his idea to send the confidential Einstein-Szilárd letter in August 1939 to Franklin D. Roosevelt outlining the possibility of nuclear weapons. The two brilliant and influential Jewish scientists feared the irresistible rise of a Nazi Bomb.

Within months the Manhattan Project was launched. It would ultimately boast over 130,000 employees, a total cost of $2 billion ($26 billion in 2015 figures), and the detonation of three nuclear weapons in 1945: The Trinity test detonation in July in New Mexico; a uranium bomb, "Little Boy," detonated on August 6 over Hiroshima; and a plutonium bomb, "Fat Man," discharged on August 9 over Nagasaki.

As the war raged on, Szilárd became sickened that scientists were losing power over their research. The military maneuvers were sinister. He had hoped that the US government, resolutely opposed to the bombing of civilians prior to the war, would not use nuclear weapons on civilian populations. He hoped that the mere threat of such weapons would force Germany and Japan to surrender. So Szilárd led a petition, signed by seventy Chicago scientists, urging President Truman to demonstrate the Bomb, not use it against cities as in *The World Set Free*.

But Wells's nightmare became factual terror over Japan. As the 320,000 inhabitants of Hiroshima were waking up, the Bomb airburst over the city. Within a second, thousands were slaughtered by the heat death. Vaporized by the light and energy of the blast, shadows on the walls were their only ghostly remains. They were the lucky ones. Victims further from the detonation were blinded, or had their skin and hair ablaze. Later they would lose the white blood cells needed to fight the escalating disease.

Back in Los Alamos, many of the Manhattan Project scientists had celebrated news of the Hiroshima massacre. Austrian-British physicist Otto Frisch recalled how "Somebody opened my door and shouted, 'Hiroshima has been destroyed.'" Frisch felt nothing but nausea when he saw how many of his friends were rushing to celebrate. "It seemed rather ghoulish," he thought, "to celebrate the sudden death of a hundred thousand people."

The Manhattan Project's lead scientist was Robert Oppenheimer. He spoke for many physicists when he declared, "In some sort of crude sense which no vulgarity, no humor, no overstatement can quite extinguish, the physicists have known sin; and this is a knowledge which they cannot lose." Oppenheimer believed that if atomic bombs were to be added as new weapons to the arsenals of a warring world, "Then the time will come when mankind will curse the names of Los Alamos and Hiroshima. The people of this world must unite or they will perish." Meanwhile, the Cold War had begun.

DOCTOR MANHATTAN

Through H. G. Wells, science fiction had invented the Manhattan Project and the Atomic Age. Fantastic fiction had long imagined doomsday. The strange settings for the end of the world had begun with Mary Shelley. Human creativity in science morphed into unknown power, capable of destroying our entire species. So it was with *Frankenstein*, "the first great myth of the industrial age." Victor Frankenstein wrestles with alien forces that, "might make the very existence of the species of man a condition precarious and full of terror."

Now, science fiction had been instrumental in the development of a new apocalyptic threat. It was time for critical insight. At this focal moment in history, many felt that science fiction was alone in its ability to project ways out of this predicament. It became the means by which a mass audience was confronted with the possibility of holocaust and mutually assured destruction. No other literature came close.

It was the same story with Alan Moore's creation of Doctor Manhattan. Moore's pedigree was beyond dispute. The British writer has been dubbed a "free spirit, the best writer in the history of comic books," "the Orson Welles of comics," and "the undisputed high priest of the medium, whose every word is seized upon like a message from the ether." And by 1986, the year Moore came to create his graphic novel miniseries, *Watchmen*, and Doctor Manhattan along with it, he had already contributed to *Doctor Who*, *Star Wars*, *Marvel Superheroes*, *The Daredevils*, and *2000 AD*.

The 1980s was a paranoid period. The world seemed more dangerous than ever. Weapons of mass destruction, atomic and nuclear, were woven into the fabric of everyday life: hiding in the backseat of your head, sub-texting many news reports from around the globe, when a plane passed loudly overhead, or when your local authority tested the area's sirens. And into this climate, Moore set out to construct the character of Doctor Manhattan.

Moore was well aware of the role of science in making the Bomb. Many believed now that man would only continue to play a part in history by waging war on his innate evil. Responsible, moral actions of scientists trigger a series of consequences, for good or ill. The Manhattan Project was viewed as the ultimate abdication of such moral responsibility, of which scientists were directly accused. True, a minority of scientists were strongly motivated to communicate the dangers of nuclear war. Some, like Leo Szilárd, even dedicated the rest of their lives to this effort.

But in other physicists the guilt ran deep. On a visit to Los Alamos, Austrian journalist, Robert Jungk, had told of his encounter with a mathematician. "His face was wreathed in a smile of almost angelic beauty. He looked as if his gaze was fixed upon the world of harmonies. But in fact he told me later that he was thinking about a mathematical problem whose solution was essential to the construction of a new type of H-Bomb."

For this true-life scientist, who never observed a single explosion of the bombs he helped detonate, "Research for nuclear weapons was just pure mathematics, untrammeled by blood, poison, or destruction." As Italian physicist Enrico Fermi famously said, "Don't bother me about your scruples. After all, the thing is beautiful physics."

English physicist Freeman Dyson had suggested that "scientists rather than generals took the initiative in getting nuclear weapons programs started," and that they were "motivated to build weapons by feelings of professional pride as well as of patriotic duty," rather than strategic needs. And Philip M. Stern wrote, "If scientists as sensitive as Oppenheimer can indeed wall off their moral sensibilities so completely and successfully, then technology is an even more fearsome monster than most of us realize."

Doctor Manhattan was partly based on DC Comics' Captain Atom, whom Moore had originally meant to portray as being surrounded by the shadow of nuclear threat. But Moore found the complexity and nuanced character of Manhattan, "a supreme super-hero," would be far more fitting a comparison with the likes of Oppenheimer than Captain Atom ever would.

Oppenheimer had said that "when you see something that is technically sweet you go ahead and do it . . . that is the way it was with the atomic bomb. I do not think anybody opposed making it; there were some debates about what to do with it after it was made." Polish-British physicist and nuclear skeptic Joseph Rotblat had agreed. "Scientists felt that only after the test . . . should they enter into the debate about the use of the Bomb." And even Freeman Dyson acknowledged the draw. "I have felt it myself, the glitter of nuclear weapons. It is irresistible if you come to them as a scientist. To feel it's there in your hands—to release this energy that fuels the stars—to lift a million tons of rock into the sky," he felt was, "partly responsible for all our troubles."

In making Doctor Manhattan, Moore wanted to delve into nuclear and quantum physics. Moore thought that a character, living in such a quantum cosmos, wouldn't see space and time with the same linear perspective as the rest of us. Manhattan would have a totally different perception of human affairs. And Moore had in mind Spock from *Star Trek*. But he didn't want the immediate alienated otherness of Spock. Rather, he wanted

Doctor Manhattan to keep his human habits, but to slowly evolve away from them, and humanity in general, which is perfectly encapsulated in Manhattan saying, "The photograph is in my hand. It is the photograph of a man and a woman. They are at an amusement park, in 1959 . . . I'm tired of looking at the photograph now. I open my fingers. It falls to the sand at my feet. I am going to look at the stars. They are so far away, and their light takes so long to reach us . . . All we ever see of stars are their old photographs . . . It's October, 1985. I'm basking in the two-million-year-old light of Andromeda. I can see the supernova that Ernst Hartwig discovered in 1885, a century ago. It scintillates, a wink intended for the Trilobites, all long dead. Supernovas are where gold forms; the only place. All gold comes from supernovas."

MANHATTAN PORTRAIT

The glowing, blue figure of Doctor Manhattan in the 2009 *Watchmen* movie is one of the true stand-out super-beings of the genre.

So how does Manhattan's emergence from the history of nuclear physics come out in his character? Many of Manhattan's powers have quantum aspects because of, according to Moore's acclaimed graphic novel, Doctor Manhattan's accident in a chamber called an "*intrinsic field subtractor.*" Fantastic fiction very often requires a "willing suspension of disbelief." The poet Samuel Taylor Coleridge first coined the phrase, in 1817. Coleridge suggested that if a writer could infuse a "human interest and a semblance of truth" into a fantastic tale, the reader would suspend judgement about the implausibility of the narrative.

And, with Doctor Manhattan, we need large helpings of that suspension. Quantum physics deals with the behavior of nature's elementary particles, both separately, and in groups. It explains the things that atoms and electrons might do. But it takes a major leap of faith to extrapolate quantum theory to Manhattan, a quantum-based superhero who can, among other things, exist outside space and time, teleport, and split into different copies of himself.

Take, for example, the reasons why Doctor Manhattan might be blue. Manhattan's blue color in the original graphic novel was at first an aesthetic choice of the artist, who had also created the blue character Rogue Trooper. A blue skin motif for Manhattan was used as it resembles skin tonally, but also served to make Manhattan's color scheme unique. But the blue can also be related to quantum science. Consider Cherenkov radiation. Named after 1958 Nobel Prize winning Soviet scientist, Pavel Alekseyevich Cherenkov, Cherenkov radiation is an electromagnetic radiation emitted by particles moving in a medium at speeds faster than that of light in the same medium.

As a result of his molecular disintegration and reconstruction, Manhattan could be leaking high-energy electrons. Since he had to rebuild himself, atom-by-atom, he's likely to have all manner of loose electrons, flying off at great speeds, and giving him that blue glow. In fact, if Manhattan changed the speed of those electrons somehow, he'd even change how deep a blue he was, as he does at times in the graphic novel.

And what of the physics of Manhattan's transfer from human to super-being? Is it possible to annul the forces holding someone's atoms and nuclei together, so they disintegrate? Consider the four forces of nature. As you sit reading this book, whether in paper form or e-book, you may not be entirely aware of the forces acting upon you. Now, a force is the influence that changes movement. A push or a pull that changes an object's motion, or makes an object deform.

You'll no doubt know the familiar force of gravity. It appears to pull you down into your seat, toward the core of the planet. You feel it as your weight. But how come you don't fall right through your seat? Electromagnetism. That's the force, which holds the atoms of your seat together, stopping *your* atoms from invading those of your seat. The electromagnetic interactions in your computer are to blame for the light that allows you to read the screen.

Whereas the forces of gravity and electromagnetism are reasonably easy to observe every day, the other two of the four fundamental forces of nature are not. That's because they work at the atomic level, the realm of Doctor Manhattan. The third force is the strong force, which holds the nucleus together. And, lastly, the weak force is responsible for radioactive decay, particularly beta decay, where a neutron within the nucleus changes into a proton and an electron, which is ejected from the nucleus.

How does the power of the four forces compare? To answer that question, consider the force you are most used to—gravity. The gravitational force is incredibly weak. So much so that the weak nuclear force is 10^{28}—ten billion, billion, billion—times stronger than gravity. Electromagnetism is 100 billion times stronger again, and the strong nuclear force (the clue is in the name) is 100 times stronger still. American science writer Timothy Ferris compares the four forces through the ingenious scientific device of a toy poodle. If the poodle's shortest leg, at one inch long, were to represent gravity, then the leg representing the strong nuclear force would be far longer than the radius of the observable Universe. That poodle has some limp.

Scientists sometimes imagine a force that is a unification of the strong, weak, and the electromagnetic forces. But that would need energy so vast it's hard to imagine creating it, unless you're Alan Moore. It would need a particle accelerator a trillion times more powerful than the most powerful particle accelerator ever built.

But if we comply with Coleridge and willingly suspend our disbelief, imagine we could annul the forces that hold us together as humans. There'd be no electromagnetism binding your atoms. There'd be no strong force holding fast your nuclei. The "intrinsic field subtractor" really would disintegrate you at the subatomic level.

When all this happened to Jonathan Osterman, his body was disintegrated, but his consciousness survived. And that meant the good Doctor was not only able to rebuild himself, he went one step further to add a six pack, broad muscular shoulders, and the kind of generally sculpted form that is exceedingly rare in the field of physics!

Finally, what did our quantum superhero have to say about nuclear annihilation? At first his condition distances him somewhat from human affairs, and he forgets that he too was once human. "I am tired of this world, these people. I am tired of being caught in the tangle of their lives." And when asked, "Humanity is about to become extinct. Doesn't it bother you?" Manhattan replies, "All that pain and conflict done with? All that needless suffering over at last? No, that doesn't bother me." And to "Jon, what about the war? You've got to prevent it! Everyone will die," he replies, "And the universe will not even notice." Echoing his view that human life is, "a highly overrated phenomenon. Mars gets along perfectly without so much as a micro-organism."

In the end, however, when the world actually stands on the brink of nuclear annihilation—a threat no less remote today—Doctor Manhattan finally feels that human life is a miracle worth saving. "In each human coupling, a thousand million sperm vie for a single egg. Multiply those odds by countless generations, against the odds of your ancestors being alive, meeting, siring this precise son; that exact daughter . . . until your mother loves a man . . . and of that union, of the thousand million children competing for fertilization, it was you, only you . . . (it's) like turning air to gold . . . thermodynamic miracle."

ARE WE ALL EVOLVING TO BECOME LIKE THE JUSTICE LEAGUE'S CYBORG?

"My body may have its limitations, but when I put my mind to it, there's nothing I can't do."
—Cyborg, *Teen Titans, Season 2: Only Human* (2004)

"Some seem unaccepting in this transformation, and it indeed has been gradual. In a sense, it began when the first simple machines were invented. But now, to deny the change requires a willful ignorance since, if you observe bodies clothed in steel flowing over highways, or how we've outsourced half our memory to these devices, these exo-brains we carry around, and if you note how even our most intimate relationships occur remotely, at great distances from one another, if you see all this, well, it isn't such an original observation, dear cyborgs, to say that human and machine long ago merged inextricably."
—Eugene Lim, *Dear Cyborgs* (2017)

"We are programmed to be dissatisfied. Even when humans gain pleasure and achievements it is not enough. They want more and more. I think it is likely in the next 200 years or so homo sapiens will upgrade themselves into some idea of a divine being, either through biological manipulation or genetic engineering of by the creation of cyborgs, part organic, part non-organic. It will be the greatest evolution in biology since the appearance of life. Nothing really has changed in four billion years biologically speaking. But we will be as different from today's humans as chimps are now from us."
—Yuval Noah Harari, the *Daily Telegraph* (2015)

"Just once I'd like to see something where cyborgs aren't the enemy. Cybermen, Cylons, and Borg, oh my. Do you have any idea what that kind of negative representation can do to your self-esteem?"
—Victor Stone, *Cyborg Vol 1, #5* (2016)

Some call him cyborg. And yet he's also been known as Cyberion, Robotman, Technis, Bionic Man, Cyborg 2.0, Omegadrome, Sparky, The Man with the Iron Fists, Tin-Man, and Silver Fists. To say that Victor Stone grew up with a few problems is not really doing justice to his case. His folks, Silas and Elinore Stone, were scientists who used young Vic as a test subject for their many intelligence enhancement projects. It's hardly surprising that Victor resented being a boy wonder guinea pig, striking up friendships with other young miscreants and running into trouble with the law. And yet at the same time Victor is heading down a dark path, which often finds him injured, his parents' experiments prove ultimately successful, with Vic's intelligence shooting up to genius levels, and an IQ measured at 170.

How enhanced is Victor? With a good deal of his body replaced or augmented by advanced tech, Vic is so enhanced that his strength, speed, stamina, and flight are all superhuman. The metallic nature of his tech means that he's a lot more durable than normal humans. And given his onboard computer system can interface with external computers, the "cyber" part of Victor's identity is pretty much maxed out. And if that's not enough, he has superhuman vision due to his electronic "eye," a wide range of tools and weapons, such as a grappling hook and a finger-mounted laser, and maybe his most often-used weapon: his sound amplifier, a "sonic cannon," which allows Victor to stun the ears of his enemies, or to pulse out focused blasts of sound, potent enough to dent metal, or shatter rock.

What's more, Victor is an *evolving* cyborg. He's been able to tinker over time, tweaking his tech parts, refining his functions and abilities to levels even beyond those set by his folks. So, for example, Victor's self-repair system is far beyond the mass production cyborg version built by fictional industry. This system enables him to seamlessly repair his body tech, no matter how tattered or torn, and even enhance the health of his biological parts to some degree. The revamped Victor, in DC Comics' 2011 re-launch,

The New 52, takes his tech a little further. As well as the abilities above, Victor now has program adapters, which allow him to interface with other body extension tech. He can fly via rocket boosters, and use jump jets to leap great distances. He's worked with Batman to create a better data front-end to interface with the worldwide networking system, and his now cybernetic lungs enable him to breathe underwater, like Aquaman. But are humans likely to become cyborg, as we gradually fuse our tech, such as computers and smart phones, with our bodies?

THE CYBORG-ARMY BEGINS AT HOME

Do you have any cyborgs in your family? I'm not talking about your grandpa in glasses, though admittedly he may well be a mean fusion of man and machine. Contact lenses and hearing aids appear not to count either, but at least they represent some blurring of that ages-old man-machine interface. In the history of science fiction and science, cybernetics stands for a combination of *cybernetics* and *organism*. Organism is any living thing. Whereas cybernetics is the scientific study of how humans (or aliens, animals, and machines) control and communicate data. According to this definition, a cyborg can be anyone with a heart pacemaker, an artificial joint, an insulin pump, or a cochlear implant. So you may well have a cyborg in your midst.

One of the more unlikely early examples of a cyborg in fiction is the Tin Man in L. Frank Baum's 1900 classic book *The Wizard of Oz*, along with the very many movie versions that followed. The Tin Man was Dorothy's hero and companion on the Yellow Brick Road. The most famous film version of *The Wizard of Oz* was made in 1939 and the associated movie, *Oz, the Great and Powerful*, in 2013.

What makes the Tin Man a cyborg? In the original tale, he was a lumberjack called Nick Chopper (not my joke, for once). He was engaged to a munchkin girl named Nimmie Amee (you just can't make this stuff up; actually, you *can* make it up, as Baum clearly did). Anyhow, the Wicked Witch of the East conjures up a magic axe that chops off his limbs one by one (these old fairy tales are often very dark). He gradually gets them replaced with metal versions and becomes a type of early cyborg.

The book that made cyborgs famous in the US was a novel called, unsurprisingly, *Cyborg*, written in 1972 by Martin Caidan. Most folk will know *Cyborg* for the name of its hero, Steve Austin. Austin was a fictional test pilot who had a near-fatal air crash and had large parts of his body replaced with bionic limbs. For most of the mid-1970s, Steve Austin was famous on American television as *The Six Million Dollar Man*, a reference to how much it cost to rebuild him into a cyborg.

Since then, cyborgs have become something of a science fiction staple. And, like the ruthless Darth Vader and the demonic Daleks in *Doctor Who*, cyborgs are among science fiction's most famous villains. Indeed, *Doctor Who* has also produced one of the most prominent of cyborg incarnations in the form of the Cybermen.

Appearing first on British television way back in 1966, the Cybermen were a fictional race of cyborgs, a totally organic species to start with, who began to implant more and more artificial parts to help them survive. They are said to originate from Earth's twin planet Mondas (this is totally made up, as we don't actually have a twin planet, last time we looked). As the Cybermen, like Victor Stone, added more and more cyber parts, they became more coldly logical, calculating, and less human. As every emotion is deleted from their minds, they become less man, and more machine. Quite ingenious, really, as you can also use the idea of Cybermen as what humans may one day become, if we base all our decisions on cold calculation, and ignore our more human and emotional aspects.

SCHOLARS PLAY CATCH-UP

Scientists and engineers weren't far behind, as science eventually caught up with fiction. In 2000, Dr. Miguel Nicolelis, a neurobiologist at the American Duke University Medical Centre, taught a monkey to use a robotic arm. The cheeky monkey's thoughts were transmitted to the arm using electrodes that were planted in the monkey's brain. This test case showed how bionic limbs could be controlled, as the idea of real life cyborgs gained ground.

Then, in 2015, Yuval Noah Harari, a professor at the Hebrew University of Jerusalem, said the near-future cyborg fusion of man and machine would soon become the "biggest evolution in biology" since the emergence of life on Earth, four billion years ago. Prof Harari had hit the headlines in 2014 with his book, *Sapiens: A Brief History of Humankind.* Originally published in Hebrew under the title *A Brief History of Mankind*, Harari's work had become a global phenomenon, attracting a legion of fans from Bill Gates and Barack Obama to Chris Evans and Jarvis Cocker, and published in nearly forty languages worldwide.

Now, Harari used the expertise gained from charting the history of humanity and turned his eyes to the near future. His verdict? Mankind would evolve to become like gods, with power over death, and become as altered from the humans of today as we are from apes, maybe more so. Central to Harari's argument is the idea that the human race is a striving species. We are driven by dissatisfaction, and we simply won't be able to resist the urge to "upgrade" ourselves, whether it's by genetic tinkering or augmenting tech.

But Harari's cyborg future is not for everyone. Given the potential expense of the tech, he feels it's likely the imminent "go cyborg" option will be limited to the rich. And while the wealthy have the money to "go all Batman" and potentially live forever, the poor will simply die out. What had led to humanity's dominance, now and in the future? According to Harari it is the human ability to invent "fictions," which hold society intact—fictions such as money, religion, and the idea of basic human rights, all of which have no basis in nature. "What enables humans to cooperate flexibly, and exist in large societies is our imagination. With religion, it's easy to understand. You can't convince a chimpanzee to give you a banana with the promise it will get twenty more bananas in chimpanzee heaven. It won't do it. But humans will. Most legal systems are based on human rights but it is all in our imagination. Money is the most successful story ever. You have the master storytellers, the bankers, the finance ministers telling you that money is worth something. It isn't. Try giving money to a chimp. It's worthless. God is extremely important because without religious myth you can't create society. Religion is the most important invention of humans. As long as humans believed they relied more and more on these gods they were controllable. The most interesting place in the world from a religious perspective is not the Middle East, it's Silicon

Valley where they are developing a techno-religion. They believe even death is just a technological problem to be solved. But what we see in the last few centuries is humans becoming more powerful and they no longer need the crutches of the Gods. Now we are saying we do not need God, just technology."

WHAT'S NEEDED TO DEVELOP REAL SUPERHERO TECH?

Stark Industries: *"Changing the World for a Better Future"*; Wayne Enterprises: *"Look What WE Can do"*; LexCorp: *"Let's Build a Better Tomorrow, Today"*; and Oscorp: *"Altering the future, from the cell to the superstructure"*

—Various slogans of superhero private capital

"One of the benefits of space exploration is international cooperation. Although the Age of Space began in a fiercely competitive mode, political and funding realities have now shifted the balance toward cooperation."

—NASA, quoted in *Why We Explore* (2005)

"Once you have an innovation culture, even those who are not scientists or engineers—poets, actors, journalists—they, as communities, embrace the meaning of what it is to be scientifically literate. They embrace the concept of an innovation culture. They vote in ways that promote it. They don't fight science and they don't fight technology."

—Neil deGrasse Tyson, in *Forbes: Intelligent Investing* (2012)

"No country, no business class, has ever been willing to subject itself to the free market, free market discipline... [The] enlightened states... resorted to massive state intervention to protect private power, and still do ...Virtually the entire dynamic economy in the United States is based crucially on state initiative and intervention: computers, the internet, telecommunication, automation, pharmaceutical, you just name it."

—Noam Chomsky, *Sovereignty and World Order address* (1999)

Would you trust the development of superhero tech to the kind of lone-genius mad doctors you get in superhero stories? Spiderman had to deal with Doc Oc, a.k.a. Doctor Otto "Octopus" Octavius. The quintessential mad scientist, Doc Oc developed a set of advanced mechanical arms, controlled by a brain-computer interface. Though essentially a cyborg, Doc Oc got his rather derogatory name from co-workers who despised him, though in the 2004 *Spiderman 2* movie, he's christened Doctor Octopus by the New York City tabloid newspaper, *The Daily Bugle*. Obsessed with proving his genius, Doc Oc works out of his laboratorial refuge at the harbor, funded by bank robbing.

One of Batman's greatest bugbear scientists was Victor Fries, of course. Also known as Mr. Freeze, Victor not only shared a first name with Frankenstein, but also cut a similarly tragic figure in the DC Universe. No matter what permutation of the Fries backstory you buy into, Victor, with his refrigerated suit and his "Freeze Ray" ice gun, remains one of the best mad scientists in comics.

Finally, there was Superman's archnemesis, Lex Luthor. Luthor is conducting an all-out science war against the Man of Steel. Though it's often billed as a mere brain vs. brawn story, the truth is that Luthor draws on the lone mad scientist aspect of his character in the continuing clash. It's a reminder to all of us that, though Superman may be an alien demigod, driven by moral integrity and responsibility, Luthor is a threat as he has the genius of science on his side. He uses his intellect, and even builds a war suit to match Superman's Kryptonian solar-strength in battle.

THE MAD DOCTOR CULTURE

Where do these mad doctor depictions come from? Think about the Golden Age of Superhero Comics, between about 1938 and 1950. The culture into which such superhero tales came was already full of the mad doctors of fiction and film. It had started way back in an Elizabethan 1588, when Christopher Marlowe's play, *Doctor Faustus*, had first been performed. Faustus's claim to mad doctor fame is that he had negotiated a pact between himself and the Devil. Faustus had been a learned scholar, who in his quest for the true essence of life summons Mephistopheles (the Devil), who offers to serve him as long as Faustus lives. The play was based on dodgy and dubious German magician and alchemist, Doctor Johann Georg Faust.

And, with Mary Shelley's 1818 novel, Faustus took new form as Victor Frankenstein, the first great myth of the industrial age. Frankenstein is in many ways the original and still the maddest of mad doctors. Importantly, Frankenstein was, of course, the name of the mad doctor, not the creature. The subtitle of Shelley's book had been, *Frankenstein; or, The Modern Prometheus*. It begs comparison with Prometheus's theft of fire from the gods for humanity's profit, with Victor Frankenstein's dream of unlimited power through science. And so all the mad doctor stories begin here, with Frankenstein's simple message: sole scientist as God, conjures some creation, and chaos ensues.

By the dawn of the twentieth century, these mad doctors had found their way into moving pictures. As early as 1910, cinema brought Frankenstein to the silver screen. The Thomas A. Edison Studios produced a sixteen-minute trick film, the first ever adaptation of Mary Shelley's definitive mad doctor book. In a move from the horror elements of Shelley's story to the more science fictional, the studio press kit for the movie said it had removed all the tale's "repulsive situations," and focused instead on the story's "mystic and psychological" elements. And so Shelley's tale of a creature composed from corpses found in graveyards and charnel houses was jettisoned in favor of the kind of alchemical characters that later became common in superhero comics. The Edison Studios press kit also explains one of the most celebrated sequences in the short film, which greatly influenced the comic books to come, and in which you can also quite easily imagine *Gotham*'s Hugo Strange: "Alone in his room [Frankenstein] conducts the experiment and after an almost breathless suspense is rewarded by seeing an object forming and rising from the blazing cauldron in which he has poured his ingredients— a vague, shapeless thing at first, but which gradually assumes a human form and exhibits signs of animation. His joy at the success of his experiment is quickly turned to horror and dismay."

Arguably even more influential than Frankenstein was Fritz Lang's *Metropolis*. This 1927 sci-fi cinematic extravaganza, produced in Germany at the height of the Weimar Republic, was the most expensive silent movie of its day. Fritz Lang's stylistic and seminal work has been dubbed "Raygun Gothic." The film featured an architecture based on contemporary

Modernism and Art Deco. Its futuristic dystopia of skyscrapers and social class is only a short stone's throw away from the typical portrayal of Gotham.

And set in the middle of this Raygun Gothic was the evil genius Rotwang, the most significant mad doctor in the history of the movies. The title-card of *Metropolis* read, "In the middle of the city was an old house," an early announcement of Fritz Lang's central belief that "an audience learns more about a character from detail and décor, in the way the light falls in a room, than from pages of dialogue." Is that not a perfect recipe for the mad doctors portrayed later in superhero graphic novels? The novel *Metropolis*, upon which the screenplay was based, gives more details about the gothic monster at the heart of the science of such mad doctors, "There was a house in the great Metropolis which was older than the town. Many said that it was older, even, than the cathedral . . . Set into the black wood of the door stood, copper-red, mysterious, the seal of Solomon, the pentagram . . . It was said that a magician, who came from the East (and in the track of whom the plague wandered) had built the house in seven nights . . . Parchments and folios lay about, open, under a covering of dust, like silver-grey velvet . . . Then came a time which pulled down antiquities . . . but the house was stronger than the words on it, was stronger than the centuries. It hardly reached knee-high to the house-giants, which stood near it. To the cleanly town, which knew neither smoke nor soot, it was a blot and an annoyance. But it remained."

MAD DOCTORS GO CORPORATE

What do we have, in terms of the defining influences of the development of superhero tech? A gaggle of lone-genius mad doctors, working mostly in isolation. And the depiction of these nutty professors belongs to another long-gone day: the original mad doctor in Frankenstein, based partly on an ages-old Elizabethan tale, and the evil genius Rotwang, who's essentially gothic origins are at the very heart of his dubious science. Even when superhero stories moved on from these dark origins of a singular maverick scientist, private capital was almost always the new king in town: Wayne Enterprises or Stark Industries, LexCorp or Oscorp, Pym Tech or Roxxon.

Let's take a closer look at their company portfolios. Wayne Enterprises, a company that sells pretty much everything, run by an absentee CEO, who battles with Gotham crime, dressed as a flying mammal. Stark Industries, a corporate brand that took a market mauling when its battery-powered CEO involuntarily created an army of murderous bots, which raised an entire European city into the sky. LexCorp, a company with huge stocks of radioactive kryptonite, and whose CEO is maniacally and fetishistically fixated on the destruction of a single alien. And finally, Pym Technologies, a brand that is far too reliant on the market viability of ant-talking tech, and Roxxon, many of whose former CEOs languish in jail. Not to mention Queen Consolidated, Veidt Enterprises, S.T.A.R. Labs, Vistacorp, Ace Chemicals, Janus Cosmetics, or Hammer Industries.

Maybe we should not be so surprised at the comic books' monstrous portrayal of science in action. After all, the modern science upon which the stories are based is itself replete with plenty of potential monsters: the thermonuclear bomb and nuclear stockpiles; continuous surveillance and the killer computer; the unravelling human genome and a new Frankenstein century; the deadly strain in a shrinking world; the rise of the robots and the prospect of enforced leisure; the macabre brainchild, rising up against its inventor; school-kids on psychotropic meds and the unchallenged power of Big Pharma.

For many scientists, these perversions of their "genius" have been laid unjustly at their door. They are the bastards of technology. Sins of applied, not pure, science. But for many writers of comic book stories, this is a much-muddled division of labor. Science is guilty, and stands firmly in the dock. The pictures portrayed in books and on the silver screen resonate strongly with reality. And mad doctors of the movies, from Doctors Faustus, Frankenstein, and Jekyll, to Doctors Strangelove and Cyclops, tell us much of the age in which we live.

Across the world, the brave new future heralded by the dawn of a technological twenty-first century is starting to show its first major cracks. At its heart lies a crisis in the public perception of science. For just when scientists were beginning to redeem their image by reinventing themselves as "green," media portrayals drag them back down to the immoral depths of Faustus and Frankenstein.

So, here we are today, still waiting for private capital to deliver our supertech future. The future we seemed to be promised in comic books. The future where we live in a world of silver flame-retardant jump suits, ray guns, and x-ray specs. You know that future, the one in which we're also no doubt meant to be invisible and immortal by now. That science fiction future of cool technologies, in which everyone zooms around in flying cars. It's summed up by the impatience of questions such as, "Dude, where's my jetpack?"

Perhaps the answer is this: We, the people, need to take back control of the project. We demand our jetpacks! Maybe the best way to get supertech for everyone, and not just the rich, is to promote more democratic government intervention. Such intervention worked in the past in the development of the general-purpose computer, the Internet, telecommunication, and automation. With CEOs often little better than flying mammals and battery-operated bots, it's interesting to note that all one hundred of the leading trans-nationals in the Fortune list have benefitted from state intervention, and twenty of the hundred had been saved from total disaster, i.e. collapse, by state bail-out. And state funding, always with sufficient funds to draw upon, has taken one of the most backward countries on the planet from near-feudalism to a global superpower whose cutting-edge technology effectively led the space race.

One last consideration is the infamous thermonuclear bomb itself. The task of building the bomb was staggeringly difficult. And yet it was achieved by cooperation: by bringing together the best international minds on the matter, and giving them the resources to do the job. The strategy accelerated the natural evolution of scientific progress enormously. Getting government to gather the egg-heads together, the sane ones, and supercharging the mission with vast material resources is capable of transforming the course of history.

WOULD ALIEN TECH BE LIKE THE VIBRANIUM IN CAPTAIN AMERICA'S SHIELD?

"To my mathematical brain, the numbers alone make thinking about aliens perfectly rational. The real challenge is working out what aliens might actually be like. We only have to look at ourselves to see how intelligent life might develop into something we wouldn't want to meet. I imagine they might exist in massive ships, having used up all the resources from their home planet. Such advanced aliens would perhaps become nomads, looking to conquer and colonise whatever planets they can reach. If aliens ever visit us, I think the outcome would be much as when Christopher Columbus first landed in America, which didn't turn out very well for the American Indians."

—Stephen Hawking, the *Daily Telegraph* (2010)

"A picnic. Picture a forest, a country road, a meadow. Cars drive off the country road into the meadow, a group of young people get out carrying bottles, baskets of food, transistor radios, and cameras. They light fires, pitch tents, turn on the music. In the morning they leave. The animals, birds, and insects that watched in horror through the long night creep out from their hiding places. And what do they see? Old spark plugs and old filters strewn around . . . Rags, burnt-out bulbs, and a monkey wrench left behind . . . And of course, the usual mess—apple cores, candy wrappers, charred remains of the campfire, cans, bottles, somebody's handkerchief, somebody's penknife, torn newspapers, coins, faded flowers picked in another meadow."

—Arkady and Boris Strugatsky, *Roadside Picnic* (1972)

"To expect alien technology to be just a few decades ahead of ours is too incredible to be taken seriously."

—Paul Davies, *Lost at Sea: The Jon Ronson Mysteries (2015)*

"Vibranium. The most versatile substance on the planet, and they used it to make a frisbee."

—Ultron, *Avengers: Age of Ultron (2015)*

Imagine aliens visited Earth. This visit would not be like the one in *War of the Worlds*, or any of the other countless film or fiction invasion stories. In these scenarios you usually get bug-eyed monsters arriving either to wipe out and displace humans, enslave us all under some colonial system, harvest humans for food, plunder our planet's resources, or destroy our world wholesale.

Nor would this visit be like the overly ambitious, but amusing, alien invasion in Douglas Adams's *Hitchhiker's Guide to the Galaxy*, "For thousands [of] years the mighty ships tore across the empty wastes of space and finally dived screaming on to the first planet they came across—which happened to be the Earth—where due to a terrible miscalculation of scale the entire battle fleet was accidentally swallowed by a small dog."

Our alien invasion is just as imaginative. The invading aliens don't even realize Earth is inhabited. After all, it hardly looks like a significant planet, and there are no obvious signs of intelligence here. Our alien visitors simply stay the one night and then zoom off to somewhere more important and interesting. The aliens essentially picnic on our planet then depart, leaving in their wake some extraterrestrial trash, which totally transforms our world. Think this is all a bit far-fetched and fantastic? Well, such an invasion is exactly what happens in *Roadside Picnic*, a book written by Soviet science fiction authors, Arkady and Boris Strugatsky, in 1972.

Roadside Picnic is an alien invasion tale with a twist. The story is set in post-visitation planet Earth, where there are now six mysterious Zones. These regions of our globe have been touched in some way by the alien visitation, some ten years past. The Visitors were never seen. But people local to the Zones reported loud explosions that blinded some and caused others to catch a kind of plague. Though the visit is thought to have been brief, less than a day, the half dozen Zones are full of mysterious phenomena, where strange events continue to occur.

Deep in these Zones, the laws of physics break down. Some Zone districts earn ominous names from studying scientists. The "First Blind

Quarter," "Plague Quarter," and "Second Blind Quarter"—all based on effects the Visitation has had on the local population. The six Zones become contaminated with fatal phenomena, littered with mysterious objects, or alien artifacts, whose various properties and original intent is so incomprehensible and advanced it might as well be supernatural. The Visitation Zones vary in size, some located in populated towns, most perhaps the size of a few square miles, beset with abandoned buildings, railways and cars, some decaying slowly, while others seem brand new. The Zones are deadly to all forms of life. They harbor space-time anomalies, and random spots capable of killing by fire, lightning, gravity, or other bizarre and alien manners.

The location of the six Zones is not random. This discovery, made by a Nobel Prize-winning physicist, is explained in a radio interview at the beginning of the book. "Imagine that you spin a huge globe and you start firing bullets into it. The bullet holes would lie on the surface in a smooth curve. The whole point (is that) all six Visitation Zones are situated on the surface of our planet as though someone had taken six shots at Earth from a pistol located somewhere along the Earth-Deneb line. Deneb is the alpha star in Cygnus."

The governments of the world, coordinated by the United Nations, try to keep a lid on the Zones and their alien tech. The UN insists on tight control, trying to stem the flow of alien artifacts out of the Zones, in fear of the unpredictable consequences of such poorly understood alien technology. It is quite possible that a single artifact harbors sufficient power to cause a plague, permanently damage, or even destroy the planet. Pitted against this authoritarian control is a frontier culture of stalkers who inhabit the Zone perimeters. Communities of such stalkers cultivate around the Zones, a kinship of thieves whose aim is to steal into the forbidden regions and find any potentially lucrative alien artifacts. Stalkers work at night, as by day soldiers and scientists constantly observe the Zones. The story is set in the town of a fictitious country, and tells the tale of one such stalker, over a period of eight years.

The lure of the alien artifacts is hard to resist. Though many are lethal, some have beneficial powers, such as the "so-so," a round black stick that produces endless energy and is used to power vehicles. Others, such as the "Death Lamp," emit deadly rays that destroy all life in the vicinity. The holy grail of the alien artifacts is the legendary "Golden Sphere." Rumored to have enough latent power to make any wish come true, the Sphere is buried so deep in the Zone that only one stalker knows the true path to its location. Another artifact is the substance known as "witches' jelly." Referred to by scholars as a colloidal gas, the jelly penetrates most known materials. All it touches merely transforms into more witches' jelly. Only certain ceramic vessels are able to contain it. The jelly seems to collect in low-lying areas, such as basements, but is highly volatile.

ELEMENTARY, DEAR CAPTAIN

What does the story of *Roadside Picnic* tell us about the possible existence of a metal such as the Vibranium in Captain America's shield? Actually, quite a lot. Let's think about the story of the elements on Earth. Originally, at the time of the Ancient Greeks, we thought all things on our planet were made up of different amounts of the four elements—earth, air, fire, and water. The Greeks assumed these four elements were all there were in a sub-lunary sphere, which really meant the "realm of the Earth," and extended out almost as far as our Moon. Beyond the Moon, in the supra-lunary sphere, these four earthly elements are gone. The rest of the cosmos, from the sub-lunary to the sphere of the fixed stars, was made of a different fabric—the quintessence, or the fifth element. And this quintessence was powerful. It was pure, perfect, crystalline, and immutable. So, the early story was, Earth was made of the four elements, and space was simply made of different stuff.

But Italian astronomer Galileo put paid to all this mystical matter. His early experiments with the telescope showed that Heaven and Earth were made up of the same stuff. What was his evidence? Well, for one thing, Galileo pointed his scope at the Moon, and found it was all cratered and craggy, which suggested the quintessence was mutable after all. Then, when Galileo saw sunspots, he realized they were an impurity in the quintessence, which was meant to be pure and perfect. So the idea began to dawn in the minds of medieval scholars that matter was universal and mutable. In short, you'll find the same stuff anywhere in the Universe you care to look.

And you *can* look. You can even see the stuff of which distant stars are made. How? By what Richard Dawkins once called the "barcodes in the stars." Look at the light from any distant star. When you closely examine its light spectrum, you will see that it is punctuated by a series of randomly spaced black lines. They look like black and slender, silhouetted tree trunks against a rainbow background of colored light. Each line is caused by some element absorbing light in the atmosphere of the star, which is why there's an absence of color on that part of the spectrum. But how can you tell what element it is in the star, which is making the black lines? Simply by, back on Earth, making the same element glow in the lab. The energetically excited element will make a reverse pattern of bright colored lines against a black background. And from an exacting analysis of these barcodes in starlight, we can tell in great detail all the elements that sit within the stars. Not only that, but as the astronomy magazine *Sky & Telescope* put it in 1996, "To those who can read its meaning, the spectral code tells at a glance just what kind of object the star is—its color, size, and luminosity, its history and future, its peculiarities, and how it compares with the Sun and stars of all other types."

And so we know the elements from star system to star system are pretty much the same. And that includes metals. On Earth, we've known about metals for millennia. In fact, metals changed human history. The Stone Age didn't disappear overnight. Stone tools were still being used, even when humans switched from hunter-gathering to farming. And metals were a major factor in the rise of our first cities. Bronze, the alloy of copper made by joining it with tin, even gave its name to the next era of early human civilization after the Stone Age. (The word for *metal* comes from a Greek word meaning "to search," so you can see how hard this stuff was to find at first.) The change from using usually stone to mostly metal took thousands of years.

In very ancient times, only seven metals were known. These so-called Metals of Antiquity are gold, copper, silver, lead, tin, iron, and mercury. To get a good idea of how important these metals were, just think about the ancient civilizations founded upon their discovery: the ancient peoples of Mesopotamia, Egypt, Greece, and Rome.

For example, the rule of the pharaohs in Ancient Egypt conjures up gold in the mind. But gold was actually quite scarce in Ancient Egypt, and its rulers derived their power from their exclusive access to it. (The Turin Papyrus Map is an ancient Egyptian map, drawn about 1160 BC and the oldest surviving map from the ancient world, which details at least 1300 gold mines in the Eastern Desert and Nubia. In fact, Nubia even has the ancient word for gold (nub) in its name. To date, only a hundred or so mines have been found. So maybe more evidence of the ancient goldsmith's art lies beneath the sands of Egypt).

Another example of the importance of metals is the ancient Minoan civilization, which flourished from about 3650 BC to 1400 BC. The poet Homer said there were as many as ninety cities in ancient Minoan Crete. Minoan civilization shows how metal riches began to circulate through trade. The Minoans had a merchant fleet, and came to dominate the seas. They sailed for hundreds of miles, from Syria in the east, to Britain in the west, where many scholars believe the Minoans got their tin (at the time, Britain was known as "the tin islands").

"VIBRANIUM. IT'S STRONGER THAN STEEL AND A THIRD OF THE WEIGHT."

The metal of Vibranium was vitally important to the shield of Captain America. Today, there are eighty-six known metals. Vibranium doesn't rank among them. Nor is it likely to. In the Marvel Comics Universe, Vibranium is described as "a rare metallic substance of extraterrestrial origin." The source of the Vibranium in the original shield was not explained in the comics from the 1940s. But the shield's nature was detailed decades later in 2001 in a retconned tale. According to this later story, in early 1941 Captain America met with a King T'Chaka of the fictional African nation of Wakanda. The King gave the Captain a second sample of Vibranium, which was described as an "alien metal" with unique vibration absorption properties, and found only in Wakanda and the Savage Land. The new Vibranium sample was used to make Captain America's circular shield and his old triangular one was retired. The placing of Vibranium in 1940s Africa is an interesting one.

The African continent was also the focus of concern about the building of Hitler's atom bomb. In 1939, Albert Einstein had signed a letter to Franklin

D. Roosevelt. The letter had warned the President of the possibility of the Nazi regime developing an atomic bomb. In particular, the letter urged Roosevelt to protect a uranium mine in the Belgian Congo, now Democratic Republic of Congo. The mine was the richest in the world: an average of 65 percent uranium oxide, in comparison with American or Canadian ore, which contained less than 1 percent. When Roosevelt met Churchill for talks in New York in 1942 in Hyde Park, they had speculated on German progress on the Bomb. Churchill later exclaimed in horror in his memoirs, "What if the enemy should get an atomic bomb before we did! However skeptical one might feel about the assertions of scientists, much disputed among themselves and expressed in jargon incomprehensible to laymen, we could not run the moral risk of being outstripped in this awful sphere."

The Vibranium in the Captain's shield is virtually indestructible, from bombs or Banners. The Vibranium shield was tough enough to absorb the strength of Hulk. But its indestructibility wasn't just confined to terrestrial matters. Both cosmic and celestial foes also met their match. The shield repelled an attack from Thor's enigmatic hammer without much obvious damage. It was said to absorb most of the kinetic energy that was part and parcel of each impact. And that meant the Captain was truly protected, as he didn't feel the recoil or the impact forces from blocking such attacks. The physical traits of the shield also meant it could bounce off most smooth surfaces. It would be seen ricocheting many times, and with minimal loss in aerodynamic speed or stability. The shield would also absorb the impact of a fall, which allowed the Captain to land safely, even off several stories, such as the example in *Captain America: Winter Soldier*, where he escapes from the S.H.I.E.L.D.'s STRIKE squad by jumping off an elevator.

All this sounds far more like alien tech than mere earthly metal. Remember the alien artifacts of *Roadside Picnic*? Let's think about the Captain's shield in comparison. The Zones, we might compare them with Wakanda, were littered with alien objects whose various properties were so incomprehensible and advanced they might as well be supernatural. So the Captain's shield certainly qualifies as alien tech, though happily less dangerous to the wielder and far more inert. Meanwhile, what space-time anomalies, or other bizarre and alien manners, might have happened in Wakanda . . . ?

WHAT IS THE ULTIMATE SUPERWEAPON?

"That rifle on the wall of the laborer's cottage or working-class flat is the symbol of democracy. It is our job to see that it stays there."
—George Orwell, *The Complete Works of George Orwell* (1998)

"The atomic bomb made the prospect of future war unendurable. It has led us up those last few steps to the mountain pass; and beyond there is a different country."
—J. Robert Oppenheimer, quoted in *The Making of The Atomic Bomb* by Richard Rhodes (1987)

"I know not with what weapons World War III will be fought, but World War IV will be fought with sticks and stones."
—Albert Einstein in an interview with Alfred Werner, Liberal Judaism 16 (April-May 1949), Einstein Archive 30-1104, as sourced in *The New Quotable Einstein* by Alice Calaprice (2005)

"I've been wading through all this unbelievable junk and wondering if I should have given the world to the monkeys."
—Elvis Costello as God, in the song *God's Comic* on the album *Spike* (1989)

"War is God's way of teaching Americans geography."
—Jon Stewart, quoted in *2,548 Wittiest Things Anybody Ever Said* by Robert Byrne (2012)

Ancient sages knew the supreme art of war was to subdue the enemy without fighting. But no one seems to have mentioned this to the scores of fiction and fantasy writers over the centuries. A puerile obsession with superweapons abounds. The Martians in H. G. Wells's *War of the*

Worlds came to planet Earth with the intention of farming humans. Their superweapon of choice? The death ray (it's all in the name, folks). The ray's beam annihilated everything it touched: matter into flame, water into steam, human flesh into vapor.

Then there's the Death Star, that bitchin' space station in *Star Wars*. The Death Star famously blew up entire planets. *That's* how super it was. It also conveniently doubled up as a space station for mobility, but its main job was that of a superweapon. Its very appearance in the sky above your home planet was enough to make you think twice about resistance. As an instrument of fear, it's not even necessary to fire it, so maybe the Death Star heeded at least some of the advice of the ancients. Then there's the sublime, but not entirely serious, Point of View (POV) gun, from the movie version of *The Hitchhiker's Guide to the Galaxy*. The POV was wielded by angry women who were just sick of arguments that ended in shouting matches. The POV caused the male target to understand the perspective of the person shooting the gun. Sounds perfect for politicians too, but hardly a showstopper in the puerile superweapon stakes.

Comic books, naturally, are also replete with superweapons: the Infinity Gauntlet, a glove which somehow turns you into a God; a Cosmic Cube, whose contents have the power to worryingly warp reality itself; and Mjölnir, the Hammer of Thor, which is distinguished by the fact that it is one of the very few comic book weapons that is actually, well, a *weapon*.

Looking back, it seems this superweapon obsession has quite the history. Way back in 1627 there was a kind of Renaissance spin-doctor by the name of Francis Bacon. Bacon was the key prophet and publicist of the emerging philosophy of science. And he had a radiant vision to put scientific ideas into practice, which he was determined to showcase to the world. In his book, *New Atlantis*, Bacon listed a catalogue of potential wonders of future science. This included Renaissance superweapons, such as a more powerful cannon, enhanced explosives, and "wildfires burning in water, unquenchable." Bacon included them in his prospectus to appeal to the bloodthirsty political establishment of the day. Little changes, it seems. And even before Bacon there was Leonardo da Vinci. Da Vinci had also sought sponsorship partly based on his ingenious ability as a military mind.

Indeed, superweapons spring up where you least suspect them. Consider Jonathan Swift's book, *Gulliver's Travels*, written in 1726. Swift's novel was a parody of science, human nature, and a satire of the contemporary traveler's tales. But it also had a superweapon in the form of the flying island of Laputa. After pirates have taken over his ship in the South China Sea, Gulliver drifts ashore to a landmass called Balnibarbi. Hardly has he set foot on land when a "flying island" appears.

Made mainly of metal, and measuring four and a half miles in diameter, the floating island is magnetic. Buried inside the island is a six-yard long bi-polar magnet. It is held in abeyance by an intense magnetic field below the Earth's surface at Balnibarbi. The Laputian magnet may be turned so that the inhabited island can be steered with precision, horizontally or vertically, but only within the scope of Earth's magnetic field, which Swift estimates at four miles.

Swift had derived the idea for this floating island from the magnetic experiments of English physician William Gilbert. Swift's island is ambiguous in nature. It is a mini-planet, but also an advanced, and weaponized, technology. The island has a layered geology like a planet. The landing steps and observation galleries, however, suggest an airship. The attraction across the gulf of space between Laputa and Balnibarbi imply a planetary model. The problem, though, is that Gilbert had argued the Earth to be a magnet. So the decisive force in Swift's limited two-planet model would be magnetism, and not Newtonian gravity. Finally, the fact that Laputian scientists can engineer the position of the flying island greatly weakens its likeness to a planet.

So Swift's flying island is science fiction's first superweapon, although given it was Swift he was no doubt already poking fun at the entire thing. Laputa is akin to the Death Star, as it is a space colony of technologically superior people. The lofty and tyrannical Laputians are not the benevolent guardians of Steven Spielberg's *Close Encounters of the Third Kind*. They're far more like the brutal supervillain aliens of H. G. Wells's *War of the Worlds* and countless comic book characters of the future.

In the kind of graphic novel way of future fiction, Laputa dominates the country above which it soars. Like its Death Star counterpart of twentieth century film, any protest below is punished. By maneuvering Laputa, the land below can be deprived of sun and rain. A further reckoning may be exacted. Any discontents can have their towns subject to missile attack, or destroyed completely, by having Laputa itself plummet to Earth. And the ominous presence of this superweapon above the Earth is also part of Swift's satire on the inhumanity of science.

SUPERWEAPONS OF THE MECHANICAL AGE

Then came the days of what we now know as steam punk. By the nineteenth century Francis Bacon's vision of future science was realized. Progress had secured its dominion over nature. Science and technology established its authority in the clanging new workshop of the world that was Victorian Britain. "Were we required," wrote Scottish philosopher Thomas Carlyle in 1829, "to characterize this age of ours by any single epithet, we should call it the Mechanical Age."

Newton's system of the world was set free. The *philosophical* engine, the early steam engine, drove locomotives along their metal rails; the first steamships crossed the Atlantic; majestic transport magnates were building bridges and roads; telegraphs ticked intelligence from station to station; the great cotton works glowed by gas; and a clamorous arc of iron foundries and coal-mines powered this Industrial Revolution.

Newton had created a clockwork cosmos, a mechanical worldview. As the machinery began to mesh, science encroached upon all aspects of life. Progress and technology seemed inseparable. The machines of science were devised not merely to explore nature, but to exploit it. For every factual gadget, fiction spawned a thousand visions, meeting every challenge with a different invention. And one of the major challenges, of course, was the idea of waging war.

Steam punk naturally wanted its own weapons. The advance of weaponry became one of the key triggers of the imagination. In 1871, British Army general Sir George Tomkyns Chesney anonymously contributed a short story to a popular magazine. The highly influential *Battle of Dorking* was a vivid account of a supposed future invasion of England by the Germans, after a victory over France. Chesney's story popularized a paranoid political concern that the UK's armaments were behind the times. Better get bigger guns. In the new wave of popular fantasy the story inspired, future war fiction speculating about superweapons became more ambitious and visionary. In George Griffith's 1893 tale, *The Angel of the Revolution*, a world war was fought with airships and submarines, and forces were armed with unprecedentedly powerful bombs. Jules Verne's 1896 tale, *For the Flag*, featured a "fulgurator." This superweapon was a powerful explosive device with a boomerang action—a primitive precursor of the guided missile. And, more famously, in 1903 H. G. Wells wrote a story, *The Land Ironclads*, which foresaw the development of the tank, while MP Shiel's *The Yellow Danger*, anticipated the use of bacteriological warfare.

Some worried that, heaven forbid, women might get their hands on the weapons. Edward Bulwer-Lytton's *The Coming Race* (1871) is a book about supermen. Edward George Earl Bulwer-Lytton, First Baron Lytton, politician and novelist, was friend to creator of the modern British Conservative Party, Benjamin Disraeli. It was Lytton's intention to satirize both Darwinian biology and the ideals of John Stuart Mill's *the Emancipation of Women*. His book is a pseudo-scientific account of an evolved line of humans who believe they are descended not from apes but from frogs. The book also contains a somewhat inelegant gender reversal. The women are fitter, beefier, more assertive, and hairier, than the men.

His fascinating, if bizarre, tale is set in a subterranean world of well-lit caverns. It begins as the narrator, an American mining engineer, falls into an underground hollow. There he discovers a mysterious human-like race, the Vril-ya. These humanoids derive immense power from *vril*, an electromagnetic animating force which fuels air boats, mechanical wings, formidable weapons and automata. "In all service, whether in or out of doors, they make great use of automaton figures, which are so ingenious, and so pliant to the operations of *vril*, that they actually seem gifted with reason. It was scarcely possible to distinguish the figures I beheld, apparently guiding or superintending the rapid movements of vast engines, from human forms endowed with thought."

Bacon's dream has been realized by the subterraneans. The unearthing of *vril*, the "all permeating fluid," borne by strident emancipated females, has enabled the race to master nature. Gender equality has been achieved.

War eliminated through mutually assured destruction. It is a utopia made real, "the dreams of our most sanguine philanthropists."

Lytton, however, rejects this "angelical" social order. The sociable community of the Vril-ya has eliminated competition, but is barren of those "individual examples of human greatness, which adorn the annals of the upper world." Conflict and competition, misery and madness, are all innately human. Sounding a note struck more clearly in Huxley's *Brave New World*, Lytton brands the aim of "calm and innocent felicity" as vain dream. Utopia, and the displacement of human industry to *vril*, would lead only to enervation and ennui.

The book strikes one final fearful note. As suggested by the ominous title, once the more advanced Vril-ya surface from their caverns, they will take the place of man: "The more deeply I pray that ages may yet elapse before there emerge into sunlight our inevitable destroyers." Meanwhile, back in *The Mechanical Age*, an industrialist found his own utopia. Inspired by Lytton he made a fortune from a strength-giving beef extract elixir known as Bovril.

SUPERWEAPONS OF THE SUPERHERO COMICS

The unearthing of X-rays and radiation added more madness to this weaponized mix. The discovery of radioactivity and X-rays in the late nineteenth century fed into the infantile frenzy of future war fiction. In George Griffith's 1911 story, *The Lord of Labor*, a war was fought with disintegrator rays, which triggered the use of some kind of awesome ray in much fantasy ever since. And when Griffith, along with Wells, introduced the idea of atomic weapons, science fiction had very many field days. For example, Percy F. Westerman's *The War of the Wireless Waves*, written in 1923, was a typical near-future thriller, featuring an arms race, in this case the race between the British ZZ rays (not merely a single Z, but a deadlier double Z) and the undoubted menace of the German Ultra-K ray (Ultra here doing the job of the deadly double Z of the British).

And so this was the culture into which comic books came. It was a culture in which criminal scientists and supervillains armed themselves with marvelous, but deadly, rays or atomic disintegrators. So it should come as no surprise that the history of comic books and superheroes is also littered with the gadgets of superpower. The genre even has a villain in Thanos whose very rationale seems to be the search for the most powerful weapon. But, this being comics, the kind of superweapon encountered is even more puerile than its more conventional counterpart.

And so we get weapons such as the Mobius Chair, which can travel through all space, time, and dimensions, is impervious to harm, and is so powerful it's been used to pull two planets in its wake. Or consider one of the most powerful artifacts in the DC Universe in the laughably named Helmet of Fate (though admittedly this hat does belong to Doctor Fate). The wearer of the helmet gains powers that include telekinesis, flight, super-strength, telepathy, intangibility, super-speed, and enhanced intelligence. In truth, the supply of powers is practically endless, and plainly ridiculous.

But perhaps the prize for the ultimate superweapon goes to Marvel's Heart of the Universe. This is an energy source first discovered ages ago by a group of alien explorers who learned to harness the great energy of the cosmos. It becomes clear that, due to a fundamental flaw, the universe is doomed to end, and that this flaw cannot be corrected, even by the power of the Heart of the Universe. It seems the only way to repair the flaw in the universe would be to destroy the universe and re-build it.

And so, the puerile quest for more powerful weapons meets its logical conclusion in a device that destroys the very cosmos. If only they'd listened to those ancient sages in the first place and saved themselves all the bother.

PART IV
MONSTER

DAILY DIARY: ARACHNID DAYS AND NIGHTS—THE SWINGER'S LIFE OF A SPIDERMAN

"Whereas Superman is a godlike guy from another planet and Batman is this mysterious, unknowable billionaire, everyone in *Spiderman* is human and flawed."

—Rhys Ifans, a.k.a. Dr. Curt Connors/The Lizard in *The Amazing Spiderman* (2012)

Spiderman, Spiderman / Does whatever a spider can / Spins a web, any size, / Catches thieves just like flies / Look out! / Here comes the Spider-Man. / Is he strong? / Listen bud, he's got radioactive blood / Can he swing from a thread? / Take a look overhead . . . / Spiderman, Spiderman/ Friendly neighborhood Spiderman / Wealth and fame / He's ignored / Action is his reward. / To him, life is a great big bang-up / Wherever there's a hang-up/You'll find the Spiderman.

—Paul Francis Webster and Robert Harris, *Spiderman (theme song)* for the *Spiderman* cartoon (1967)

"And a lean, silent figure slowly fades into the gathering darkness, aware at last that in this world, with great power, there must also come great responsibility. And so a legend is born, and a new name is added to the roster of those who make the world of fantasy the most exciting realm of all!"

—Stan Lee, *Amazing Fantasy #15* (1962)

"It's spider season. Every year, right about now, thousands of the godless eight-legged bastards emerge from the bowels of hell (or the garden, whichever's nearest) with the sole intention of tormenting humankind."
—Charlie Brooker, *The Guardian* (2007)

It's 4 a.m. and pitch black.

Your bladder calls. Time to make a sleepy visit to the bathroom. That much is straightforward enough, but your nerves are frayed from playing first person shooter video games for the last forty-eight hours. Too much *Doom*, followed by far too much *Overwatch*, and so the mental map of your house has become a kind of live-action immersive-game living space. It's now the type of white-knuckle *Resident Evil* setting that could send the unsuspecting mind over the edge. When you assume your lavatorial throne, a spider that seems to be the size of a small rat scuttles across your naked feet. The adrenalin shoots through your veins like a turbo-charged pathogen. You react like you've been blasted in the butt by a mega-Taser. You simply can't move a single muscle, not even to scream. The initial horror of the shock clears and makes way for a second paralysis: the dawning realization that you've been bitten.

Are you now suddenly fit to become arachnid pals with Peter Parker? Big yourself up. You may well be the next superhero. After all, Stan Lee's *Marvel* superhero tales in the early 1960s were about ordinary folk like you. Stan revolutionized the genre with a large dose of reality. His characters were down to earth, everyday people living everyday lives. His stories were peopled with folk who had personal problems (though, naturally, there was little mention of obsessive video gaming), and his characters were often troubled. Parker's main problem, of course, was being bitten by a radioactive spider. The spider got caught in a radiation test and was slowly dying but, before it died, Parker got fanged and the spider's venom led to changes in Peter's DNA. He now had human versions of spider abilities such as proportionate strength and agility along with what became known as his "spider-sense." This was a sort of extra-sensory perception, an imminent expectation that lit up his nerves and warned him whenever danger approached. Parker also contracted the skill of spider grip, the ability to walk on walls, windows, and even ceilings.

You recall one witness to Parker's accident was a student by the name of Carl King. Now, Carl was wise enough to link Parker's mishap with the sudden appearance of the antics of the mysterious Spiderman. Hoping to jump onto the arachnid bandwagon, an admittedly curious concept, keen Carl hoped for similar powers to Peter by actually eating a radioactive spider. Things didn't go well for old Carl. His body began an irreversible entropic decay, mutating into a frenzied swarm of a thousand spiders, all sharing a single consciousness. Nightmare.

And so a realization dawns as you sit still and sleepily on your lavatorial throne: If you are to have a swinger's life like Spiderman, your arachnid days and nights might depend on the species of spider that fanged you. Come to think of it, Spiderman himself wasn't really so lucky in his inherited skills and abilities. If you were to sum up a generic spider, you'd have to admit they're pretty impressive creatures. They have a first-class sensor-array, an on-board construction system, a sturdily armored body, and a deadly venom-injection system. Compared to this list, even Spiderman was short changed, but the specifics of your skills would depend on the spider species you got bitten by.

SPIDER STATS

It's little wonder their skill set is impressive; spiders have evolved over almost 400 million years of existence. In that time they've not only developed some remarkable adaptations and been among the most successful carnivores, spiders have also spread over the continents and conquered almost every habitat on the planet, from the 6-inch camel spiders found in the deserts of Iraq, to the 10-inch sea spiders found in polar oceans. That success and geographical spread has led to about forty thousand different species of spider in the world, with possibly many more waiting to be uncovered. That's simply stunning when you realize there are only around four thousand different species in the whole mammal kingdom. That's ten different spider species for every one mammal species.

And so, will your bite simply enable you just to spin webs, attack prey, and walk straight up walls? Or, depending on species, would you also be able to swim like a fish, jump from tree to tree, and smite your mortal enemy? Given you've been bitten in the bathroom, surely the most likely outcome is that you've been bitten by a common house spider, a generic term for a variety of different spiders found in human dwellings the world over. If so, your new life of fighting crime might be somewhat limited.

A house spider's natural habitat is a house. There's more to these cunning scientific names than meets the eye. Male house spiders only really leave their webs to find a mate, so the one that's just run over your naked feet is likely to be sexually frustrated. (That explains the bite.) In early autumn each year, sexually mature males prowl around in search of females. It's at this time they're most noticeable to the human eye, running up drapes and falling into toilets and sinks. (They don't come up from plugholes as is often assumed.) In fact, 95 percent of spiders in your house have never been outdoors. It's one of the few creatures that can only live indoors, and it will actually die if it goes outside. Little do those non-arachnophobes realize, with their good Samaritan air, upturned glasses, and sheets of paper intent on "putting the little fella outside"—they're actually condemning the house spider to a grisly death. Think about what this means for your heart's desire to join Peter in his crime-fighting cause. As you sit there on your throne you imagine calling Peter up and putting forward your proposal. But, he replies, a house spider has bitten you, so you can't go outside—how can you fight crime, shouting through the lounge window of your house? Yes, you say, but perps, like many people, will have an innate fear of spiders. The perps will freeze at the mere sight of me. "We" spiders are one of those evolutionarily persistent ancestral hazards that humans are especially attuned to. Shrinks say the human visual system has an "ancestral mechanism," geared to instantly react to a spider threat that has persisted throughout evolutionary history. Are you sure, replies Peter, that the perps won't just, like, run away? And, if they do, how do you suggest you pursue them?

Peter has a good point. Desperately imagining what you will say next, you recall a recent TV documentary about zorbing, the sport of rolling inside an orb made of see-through plastic. Hell, zorbing is only a decade younger than Spiderman himself. Okay, sure, zorbing is usually done on slopes, and you can't expect perps to commit all crimes conveniently on a gradient. But then you remember zorbing can also be done on a level surface, which permits more rider control. You imagine putting this point to Peter, along with your innovation of peep-holes in the zorb ball plastic, through which you can shoot and sling your web, but Peter's no longer listening.

SPIDER SPECIES

In your mind's eye, Peter has become suspicious of your entire crime-fighting proposal. Are you sure you're not on uppers, he asks? Spiders on drugs behave just as weirdly as humans, says Peter. They create weird webs. With marijuana they never finish the job, with an upper like the amphetamine Benzedrine, they spin webs very swiftly but leave large holes, and with caffeine they make very haphazard webs. Seems to me, concludes Peter, you're on a combination of uppers and caffeine, as your swift proposal is full of large holes and very haphazard. Anyhow, he goes on, what makes you think the species of spider you were bitten by was a house spider?

Well, what other species possibilities are there, you ask? Do you feel any curious dietary pangs? replies Peter. And when asked to explain, Peter tells you house spiders survive for months without food or water, so you shouldn't feel the slightest bit peckish. By the way, he goes on, like many spiders the female house spider will eat the male after mating, but at least the female house spider waits for the male to die naturally first. The female redback spider on the other hand is one hundred times bigger than her mate, and when you're done inseminating her, you're expected to do a somersault into her mouth, at which point she gobbles you up.

You tell Peter you had no idea of this darker side of spider dating. It gets worse, he says. First Peter tells you that one of the effects of being bitten by a Brazilian wandering spider is an unwanted erection, then he spills the beans on getting a spider girlfriend. Peter explains there are spider species in which the males find cannibalistic females much more than non-cannibal females. For most male-killing female spiders, if they smell of a dead male, they're very likely to put off other males. But the male Pennsylvania Grass Spiders are far more likely to approach a female if she has recently killed and eaten a male. But that makes little or no evolutionary sense, you complain to Peter. It does sound mad, says Peter, but spider experts have done their research. Spiders usually just eat flies but, during their breeding season, females typically eat males.

This seems risky in terms of hunger because they're only likely to meet around three males in each breeding season. But, in lab tests, 75 percent of male spiders preferred to approach a female that had killed and eaten a male spider over a female that had just eaten a cricket. The cannibal females were also more likely after mating to produce egg cases that would hatch out, so this is where evolution kicks in, explains Peter. The males prefer cannibalistic females because females typically eat only one male, even if they have another opportunity to eat another male later on. And so, if your spider girlfriend smells of an eaten male, she's unlikely to kill and eat you too. But if she *doesn't* smell of deceased male, you may very well be next on her menu.

By now, in this imagined conversation, your whole demeanor has taken a turn for the horrified. Peter's spidey-sense picks this up over the telephone and he tries to help by saying, oh, don't worry, spiders don't really eat their prey. They dissolve their victims and then drink them, rather like a milkshake. And that is your final straw. Still sitting on your lavatorial throne, you decide to declare war on spider-kind. Undoubtedly still influenced by playing too much *Doom*, you decide to forget the War on Terror and call a War on Spiders. Your frenzied mind is convinced such a war would be 1) popular, 2) cheaper, and 3) actually winnable. Delicate souls the world over would sleep more soundly in their beds if they knew a crack anti-arachnid task force was permanently on call. They'd call at your home, even in the dead of night, and send any unsuspecting spider to kingdom come. That this hasn't already happened, you decide, is the greatest tragedy of our age. It's time to act. Quickly. Now. Before they bite us all and the world is done for.

I AM GROOT! HOW SENTIENT ARE TERRESTRIAL TREES?

Rocket: "All right, first you flick this switch, then this switch. That activates it. Then you push this button . . . which will give you five minutes to get out of there. Now, whatever you do . . . don't push this button . . . because that will set off the bomb immediately and we'll all be dead. Now, repeat back what I just said."

Groot: "I am Groot."

Rocket: "Uh-huh."

Groot: "I am Groot."

Rocket: "That's right."

Groot: "I am Groot."

Rocket: "No! No, that's the button that will kill everyone!"

—James Gunn, *Guardians of the Galaxy Vol. 2* screenplay (2017)

The force that through the green fuse drives the flower
Drives my green age; that blasts the roots of trees
Is my destroyer.
And I am dumb to tell the crooked rose
My youth is bent by the same wintry fever.

—Dylan Thomas, *The Force That Through the Green Fuse Drives the Flower* (1933)

"Consider a tree for a moment. As beautiful as trees are to look at, we don't see what goes on underground—as they grow roots. Trees must develop deep roots in order to grow strong and produce their beauty. But we don't see the roots. We just see and enjoy the beauty. In much the same way, what goes on inside of us is like the roots of a tree."

—Joyce Meyer, *The Christian Post* (2013)

Groot: "I am Groot."

Yondu: "What's that?"

Rocket: "He says, 'welcome to the frickin' Guardians of the Galaxy.' Only he didn't use *frickin'*."

—James Gunn, *Guardians of the Galaxy Vol. 2* screenplay (2017)

It's hard to break into the top ten superheroes of all time.

Public polls of the people's favorite comic book characters always return the usual suspects: Thor, Hulk, Captain America, Wolverine, Superman, Iron Man, Spiderman, and of course, Batman, the vengeance-seeking superhero who's been popular since his creation in 1939. So, when the Marvel studio first announced in 2012 that its next superhero epic was to be a tale of a little-known crew of multi-colored cosmic weirdos, few thought *Guardians of the Galaxy* would challenge the now familiar superhero hall of fame. But one of those cosmic weirdos was a talking alien tree with a vocabulary of only three words and whose character the people of the geek sphere took to their hearts, so James Gunn's movie became one of 2014's highest-grossing films at the global box office. And now, after we've witnessed a sequel in *Guardians of the Galaxy Vol. 2*, there are even rumors that Marvel is cultivating a solo spin-off for Groot in its greenhouse.

Groot, yet another creation of Stan Lee and Jack Kirby, first surfaced back in 1960. The creature concerned is a *Flora colossus* from the mysterious sounding Planet X, the capital of the branch worlds. The *Flora colossi* are tree-creatures whose language struggles with translation because of the inflexibility of their larynges. That's why they always seem to be repeating the phrase, "I am Groot." In fact, in one tale, *Annihilation: Conquest*, it became clear that Groot was not only able to speak, but also express himself with some considerable eloquence. There are a number of surprising plot lines lying in Groot's character bio. Originally, he was an alien tree monster (like a kind of badass Ent), who came to Earth looking for unsuspecting humans to swipe and study. On his home of Planet X, Groot is an heir of an ancient ennobled sap-line, which makes him royalty (to be precise, "His Divine Majesty King Groot the 23rd, Monarch of Planet X, custodian of the branch worlds, ruler of all the shades"). Interestingly, Groot also gives Spiderman nightmares, was a member of SHIELD's monster squad, has a soft-spot for furry animals (perhaps no surprise there), can create tiny versions of himself, was once a reality TV star, may be the last of his species, and represents Vin Diesel's greatest ever acting performance. And, just in case you thought Groot was the coolest of names, bear this in mind: in Dutch, Flemish, and Afrikaans, the word *groot* just means "big." So, watching this alien tree on the silver screen in those territories seems as though Groot is not concerned with naming himself, but rather boasting about his bulk.

Groot's species, the *Flora colossi*, enjoy the most cosmic of educations. The children of the species absorb the collected body of knowledge of many generations through the process of photosynthesis. This highly evolved type of teaching makes the *Flora colossi* geniuses, but what kind of genius might we find lurking in the roots and branches of our own terrestrial trees? The answer is far more than you might imagine.

THE TALK OF THE TREES

You probably think that trees just sit there in the wood and rarely, if ever, interact with one another, but forest ecologists say there is much more going on beneath the surface. Trees of a given species dynamically support each other. They form friendships and networks, and parent trees nurture their offspring. The evidence for this, forest ecologists claim, is in the way that one tree will grow so to avoid blocking other tree's light. Trees also send out chemical signal warnings to other trees when an insect attack is imminent. For instance, a tree under attack from aphids can suggest to a nearby tree that it might want to raise its defenses before the aphids reach it. In such ways, trees could be said to have emotions like fear and pain and a "language" which enables them to "talk" with one another.

Critics might say that ecologists are simply anthropomorphizing trees and plant life in general, but while there's little doubt that trees talk in a very different way from humans, ecologists use human references because a more scientific jargon might hinder understanding of what's going on with the life of trees. When ecologists say, "trees suckle their children," people find it far easier to know what's being communicated. One such forest ecologist is Suzanne Simard, of the University of British Colombia. Determined to show how trees are far more connected than we thought, Simard researched mycorrhizae, the symbiotic relationships that form between fungi and trees or plants. The fungi colonize the root system of a host tree, enhancing the tree's ability to absorb water and nutrients, while the tree provides the fungus with carbohydrates formed from photosynthesis. Simard explained in a TED talk in 2016 how she and other forest ecologists "set about an experiment, and grew mother trees with kin and stranger's seedlings. And it turns out they do recognize their own kin. Mother trees colonize their kin with bigger mycorrhizal networks. They send them more carbon below ground. They even reduce their own root competition to make elbow-room for their kids. When mother trees are injured or dying, they also send messages of wisdom on to the next generation of seedlings . . . so trees talk." This last gem of research from Simard sounds similar to the way that Groot's species, the *Flora colossi*, teach their tree kids by helping them absorb the collected body of knowledge of previous generations.

Simard's work hopes to persuade people to think differently about forests. She believes we'd all take more care about cutting down trees if we were more aware of the deep networks between their tree "families." Simard makes it clear that when trees send each other carbon through fungal threads, mycelium, it isn't random but rather a deliberate sending process. Her research shows that mother trees don't just prioritize their offspring with key nutrients, they also send water, nitrogen, phosphorous, defense signals, and allele chemicals through mycorrhizal networks. Mother trees can be connected to hundreds of trees in this way, and the goodness they pass to those trees increases seedling survival fourfold. Simard's research is crucial to conservation. If too many mother trees are cut down, she says, "the whole system collapses." The mycorrhizal networks, with its nodes and links, make it clear there's a wood-wide web in action. Fungi act as links, and trees as the nodes. The busiest nodes are what Simard calls the mother trees.

THE WOOD-WIDE WEB

Another ecologist working the wood-wide web is young British plant scientist, Merlin Sheldrake. One of Sheldrake's research haunts is Epping Forest, a strictly controlled wood first declared a royal hunting ground by Henry II way back in the twelfth century. Since 1878 the City of London Corporation has managed the forest, which today is almost entirely bounded by the M25, the infamous outer ring road that runs like a hangman's noose about the city of London. Smaller roads crisscross the forest (it is seldom more than four kilometers wide) and the wood is punctuated by about one hundred lakes and ponds, some of which are former pockmark blast holes of "doodlebug" rockets fired at London during World War II. And yet Epping Forest miraculously abides; nearly six thousand acres of greenwood magic—trees, heath, and waterways—a still-thriving example of a wood-wide web.

Sheldrake's specialism is mycorrhizal fungi and his research is helping the revolution that is transforming the way we think about trees and woods. Historically and for a great many centuries, fungi were believed to be harmful to trees and plants. Fungi were thought of as parasites that caused dysfunction and disease, but more recent research has changed all that. Ecologists have come to realize that the subtle symbiotic relationship between fungi and plants represent connection, not infection.

Let's look a little deeper into the connections that make up the wood-wide web. The fungi project gossamer-fine fungal tubes called hyphae that penetrate the soil and interlace with the tips of tree roots. At the level of the very cells that make up the trees root systems, fungi and roots merge to make what is known as mycorrhiza: a fusion of the Greek words for fungus (*mykós*) and root (*riza*). And so we have the wood-wide web: individual trees are networked by a subterranean hyphal linkage—a wonderfully intricate and orchestrated structure of life.

The symbiosis of mycorrhizal fungi and trees is now believed to be ancient, with some estimates suggesting a history around 450 million years old. And this affair of mutualism—in which both organisms benefit from their association—has great implications far beyond the boundaries of woods like Epping Forest. The enhanced understanding of the web's functions raises huge questions. Where does species begin and end? Might a forest be better imagined as a single superorganism?

Next time you think about Groot, try this thought experiment out for size. Imagine a terrestrial forest. Take your mind underground to where the subterranean symbiosis plays out. Imagine the soil as transparent so you can gaze down into the depths of this buried web-like infrastructure. Picture those fungal skeins suspended between tapering tree roots, emanating out into a network at least as complex as the cables and fibers that run data lines beneath our cities. British science fiction writer China Miéville once described the realm of fungi as "the kingdom of the gray," conjuring up the pure otherness of space, time, and species. Perhaps this is what is meant by Groot's title of "custodian of the branch worlds, ruler of all the shades."

DAILY DIARY: THE INVISIBLE MAN AND THE TROUBLES WITH BEING TRANSPARENT

"An invisible man can rule the world. Nobody will see him come, nobody will see him go. He can hear every secret. He can rob, and rape, and kill!"
—*The Invisible Man* (1933)

"All right, you fools. You've brought it on yourselves! Everything would have come right if you'd only left me alone. You've driven me near madness with your peering through the keyholes and gaping through the curtains, and now you'll suffer for it! You're crazy to know who I am, aren't you? All right! I'll show you! [the Invisible Man removes his rubber nose and goggles and throws them at his spectators]"
—*The Invisible Man* (1933)

"'An invisible man is a man of power.' He stopped for a moment to sneeze violently."
—H. G. Wells, *The Invisible Man* (1897)

From the very start, it seems, writers were well aware of the poignancy that superpowers can bring.

Famous Victorian British sci-fi writer H. G. Wells had his narrator suggest the Invisible Man may not be as powerful as he liked to think. At the very moment he affirms his power, an ill-timed sneeze beautifully reminds us of his frailties. Wells's *Invisible Man* was originally serialized in *Pearson's Weekly* in 1897 and published as a novel in the same year. The eponymous Invisible Man is Griffin, a scientist. His theory is this: If a person's refractive index is changed to precisely that of air, and his body does not absorb or reflect light, then he will become invisible. This Griffin does with the use of the simple expedient of a chemical elixir. But he cannot become visible again, which drives him mad.

What of the poignancy of the superpower of invisibility? What are the drawbacks of being an "invisible"? When asked what superpower they'd most like to have, kids often choose invisibility. And never, it seems, for good reasons rather than bad. Indeed, for kids and adults alike, the superpower of invisibility rarely seems to bring out the best of human nature. Naturally, the choice of invisibility revolves around the fact that "the invisibles" can spy on people without being seen and do whatever they wish without being caught. Kids usually choose robbing the neighborhood candy store as their main aspiration for becoming an invisible while the more ambitious adults target the local bank.

The 2006 British-American movie *The Prestige* explores the world of stage magic. It focuses on the way rival stage magicians in London at the end of the nineteenth century engaged in competitive one-upmanship to create the best stage illusions. By then, magic men and women had worked out how to use full-sized mirrors to bend light in order to create disappearing illusions.

And to avoid Griffin's insanity, modern physicists have donned invisibility cloaks rather than swallow a chemical formula. One such cloak was developed by Professor Vladimir Shalaev at Purdue University in Indiana. Shalaev is an American physicist of Russian descent known for his work in the fields of plasmonics, nanophotonics, and optical meta-materials. He developed a cloak made of tiny, angled metal needles that forced light to pass around the cloak. The wearer appears to vanish, without the Wellsian drawback of lunacy, and scientists like Shalaev have created meta-materials to send light rays around tiny, two-dimensional objects. Other scientists experiment with cameras that can film what is behind you and project the image so it looks like you're invisible from the front.

QUIT CHEATING

But scientists are cheating—none of these gimmicks achieve true invisible status. Real human invisibility is when a person is an invisible for all angles and distances while they're in motion. And true invisibility, from within, raises significant scientific problems that any ambitious experimenting scientist should seriously think through before they start tinkering with transparency.

For one thing, being a true invisible would also mean you'd have to be naked. If you really wanted to move about unseen, you'd simply have to strip off, no matter what the weather. And you wouldn't be able to take anything with you, either. Purse, smartphone, and car keys would all have to be left behind, or else you'd be detected simply because observers would see your knick-knacks floating in midair!

The invisibles also have additional clauses in their Highway Code. They have to be careful when walking the street, as motorists and other pedestrians simply can't see them, remember? When you're an invisible, unless you're careful you'll end up suffering from repeated pedestrian impacts, or worse, your invisible body could be totaled in a car wreck as you simply can't be seen by even the most observant drivers.

Then there's the olfactory dynamic and other factors of the senses to consider. Wearing strong perfume is right out, as anyone in your vicinity will get the full effect of your Calvin Klein, Gucci, or Chanel in their nostrils. And tread carefully, too, as any noise you make will be a sure giveaway that someone invisible like you is creeping about.

And though your body may remain invisible just like Griffin from the H. G. Wells story, material *on* your body will still be visible. An invisible is always wary of burger joints, as anyone accidentally spilling a mocha frappe on their skin is going to be unintentionally making part of them visible again. Getting caught in a rain or snow shower will have a similar effect, so invisibles are sure to take an umbrella with them at all times, though they may not always be able to shove it somewhere it can't be seen.

It's not just liquid that makes an invisible visible—beware of dust. Dust is often said to be made of a certain percentage of dead human skin, with folks quoting around 70 percent or 80 percent, but unless you're a molting bird or reptile (or you're the kind of creature that either stepped out of Doctor Frankenstein's laboratory or wanders the eerie island landscape of Doctor Moreau), very little of your environment is made of dead body parts. Sure, humans shed dead skin, but most of it is carried away by water when we shave or bathe, ending up not on our floors but in our sewers. Dust is far more likely to be made of soil particles or the tiny fibers from clothes made of cotton and other materials, animal dander, sand, insect waste, flour from the kitchen, and, naturally, good, old-fashioned dirt. Picasso may have said that art washes away from the soul the dust of everyday life, but your body, invisible or not, will accumulate dust. Every opening of a door or window stirs up the tiny, airborne particles that eventually settle around the house. But before that, they can stick to the moisture on your skin as you sweat, or to the tiny hairs on your skin when you are dry. So, you being invisible wouldn't worry dust. It would still land on every inch of you. On a daily basis, we don't usually notice the dust accumulate on our skin, as we simply can't see the thin dust layer, which sits on top of our skin color. And yet, being invisible would mean observers would see a humanoid glob of dust, just walking around the joint with super-dirty soles. Nice.

Being an invisible would also bring blindness. Have you ever thought about what the world might look like if you were an invisible? Well, the easy answer is zippo. Think about it: you can't see in the dark, as there's no light. To see a tree, light must hit the tree and return the light data to your eyes. As the retinas in your eyes capture the series of light nuances and reflections, your brain interprets them into the image of a tree, so it follows that if you're invisible, light would simply sail straight through or around you instead of bouncing off you for everyone to see. And *that* means the retinas in your eyes are not capturing the light. Bingo! Your brain has nothing to interpret into an image. Think about it—you can't see your reflection without a mirror to stop the light, so when you can't be seen by others, you also cannot see.

And what if, like H. G. Wells's Griffin, being an invisible remains a permanent condition? What happens when you fall ill? If you do, the doctor can't deliver a diagnosis or recommend some measure of medical treatment, simply because you can't be examined. And what if you're injured from one of those pedestrian impacts or are rescued from that car wreck? The doctor wouldn't know where to apply ointments or bandages because they cannot assess your injury. And, for that matter, you can't even diagnose yourself in the first place. Imagine you have a skin problem, whether it's a mild rash or the bubonic plague. How can you or the doctor diagnose the condition without being able to see the skin color change or inflammation?

So, it's not all fun, being an invisible. Sure, you can stop a robbery, get free travel, borrow someone's Ferrari, stare at people naked, ghost slap someone who annoys you, fix a sporting event, haunt someone's house, or even drive someone insane. But what if everyone was permanently invisible like Griffin? Our world would be the most tedious of places, with no people on the streets, nothing to see on social media, and no celebs on TV (okay, maybe this last one ain't so bad). All in all, being an invisible could be a very lonely business indeed.

WHEN WILL DEADPOOL'S HEALING SUPERPOWER BECOME REALITY?

"I didn't become Deadpool until after I left Project X—after I got this healing factor to cure my cancer. After my mom dies of cancer when I was a kid. After my dad died in a bar fight because of one of my drunken friends. After I'd been kicked out of the army. Which I'd signed up for as Wade Winston Wilson. Because that's who I am. And anyone who says differently . . . is just imagining things."

—Fabian Nicieza, *Cable & Deadpool Vol 1 #37* (2007)

VANESSA CARLYSLE: "So, am I supposed to just smile and wave you out the door?"

WADE WILSON: "Think of it like spring cleaning. Only if spring was death. God, if I had a nickel for every time I spanked it to Bernadette Peters."

VANESSA CARLYSLE: "Sounds like you do. Bernadette is not going anywhere, because you're not going anywhere. Drink."

WADE WILSON: "You're right. Cancer is only in my liver, lungs, prostate, and brain. All the things I can live without."

—Rhett Reese and Paul Wernick, *Deadpool* screenplay (2016)

"The day my father Odin banished me from Asgard, I was bitten by a vampire and had radioactive waste dumped into my eyes. To make matters worse, my mutant ability to control weather activated just as I was hit by a blast of gamma radiation. Nah, actually, I got this way by volunteering for the Weapon X program. They promised to cure my cancer. And they cured it all right, by giving me an outrageous healing factor. Then they labelled me psychotic and tossed me into a prison lab. So I escaped and became what some people might call a 'mercenary'. I prefer the title 'cleaner of the gene pool.' And I've made a lot of good friends along the way: like Arcade. He's always sending me to his amusement park."

—Deadpool in *Marvel: Ultimate Alliance* video game (2006)

If you were able to choose your very own superpower, what would it be?

Comic book fans and moviegoers about the globe must surely have, in the odd idle moment, considered this taxing question. While the Zen among us often plump for flight, and wicked kids often vie for invisibility, the trait of super-intelligence is often overlooked. And yet plenty of superheroes have intelligence in abundance, with Reed Richards, Bruce Wayne, Tony Stark, and Bruce Banner all weighing in with IQs way above genius level.

It seems that some other superpowers are rejected in face of their sheer impossibility. And yet, when you look a little deeper into the science, the superpower may not be as unlikely as it first seems. (This superpower situation is reminiscent of Lewis Carroll's conversation between Alice and the White Queen:

"There's no use trying," [Alice] said, "one can't believe impossible things."

"I daresay you haven't had much practice," said the Queen. "When I was your age, I always did it for half-an-hour a day. Why, sometimes I've believed as many as six impossible things before breakfast."

And it's also reminiscent of Douglas Adams's *Hitchhiker's Guide to the Galaxy*. One of the modes of travel in Adams's wonderful books is the infinite improbability drive. This method is a way of crossing interstellar distances in a "mere nothing of a second, without all that tedious mucking about in hyperspace.") Science fiction seems also to be driven by an infinity of improbabilities, using the seemingly impossible ideas as a form of plot propulsion.

Consider, for example, Spiderman's ability to walk up walls. Impossible? Scholars have recently found the secret behind the impossibility of geckos walking vertically up glass. Their feet (the geckos, not the scholars, though time alone will tell) are covered in half a million tiny hairs. Each of these hairs splits into hundreds more, with diameters less than the wavelength of light. These myriad hairs create a super-powerful bond between the electrons in the two surfaces, gecko and glass. Might human technology one day take advantage of this gecko power? One square centimeter of adhesive tape based on this lizard principle has already been manufactured. If enough can be made to cover a human hand, future humans could hang from ceilings if they so wished.

And what about the improbability of the power of supervillain, Poison Ivy? Pamela Lillian Isley was a Gotham City botanist obsessed with plants. Risky plant experiments transformed Pamela into Poison Ivy by placing an overdose of toxins into her blood stream which made her touch deadly. Pamela is just a stone's throw away from the Poison Dart Frog. If you think Pamela is impossible, you'll be blown away by the idea that a frog as small as half an inch in adult length can be the most poisonous animal on the planet. The bright color of its skin is part of the reason that the frog is active in the day. Most other frogs found in Central and South America come out at night, but the color of the Poison Dart Frog says to potential predators, "don't eat this; it's highly poisonous." Pamela is far better at disguise. The poisons created by the Poison Dart Frog are one of the miracles of nature. The poisons start in local leaves. Ants eat the leaves, the frogs eat the ants, and the poison is passed along the food chain. The frogs sweat the poison out of their skin. The poison can be so toxic that it has the power to kill ten people. This tiny little animal is a living chemical weapons factory. And for that reason, it is the most toxic creature on the planet, next to Pamela.

This question of believability extends to science fiction outside of the usual superhero subgenre, of course. The plot of the 2014 movie *Dawn of the Planet of the Apes* was scoffed at by one of my non-nerd family members as a ridiculous tale about "monkeys on horseback." And yet scholarly simian research continues to justify the imaginative creativity of the *Planet of the Apes* franchise. Gorillas are strong and silent members of the ape family. They may not be as vocal or flashy with their skills as chimps, bonobos, and the creatures of fiction, but they have excellent memories and often do things out of personal choice rather than simply for some kind of human reward. Koko, a female gorilla born in 1971 at the San Francisco Zoo, has mastered up to a thousand words in sign language. She's able to communicate complex emotions such as sadness and humor, cracking her own simian jokes. Koko describes herself, poignantly, as "fine animal person gorilla." Meanwhile, consider rhesus macaques. Unsurprisingly, perhaps, an experiment found that male macaques are happy to "pay" to look at pictures of the faces and bottoms of high-ranking females by forfeiting their regular reward of a glass of cherry juice. (With lower-ranking females, scholars had to bribe the macaques with an even bigger glass of juice before they would pay due attention.)

TIME HEALS ALL WOUNDS

This brings us to Deadpool. Superhero powers are often an exaggerated form of some power or trait possessed by creatures in the natural world: flight, strength, intelligence, or in Deadpool's case, an ability to heal and survive; a power that scholars are seriously studying and hoping to harness. As can be seen from the marvelous quotes above, Wade Wilson is a sarcastic and eccentric smartass, an extreme antihero rather than your typical superhero. Wade is powered up when he gets help from a creepy top-secret corporation to blast his cancer with a dangerous cure designed to activate mutant genes. (It's the same creepy company that gave Wolverine his adamantium skeleton and claws; these superheroes never learn.) In the 2016 *Deadpool* movie, Wade is thrown into a tiny chamber and his body is starved of oxygen for days on end. This "cruel to be kind" cure seems to work, and bingo, Wade suddenly possesses awesome regenerative powers. He can swiftly heal from any wound (even decapitation!), and his body becomes immune to most types of poisons and diseases, though the cure has left his body riddled with horrible scars.

Deadpool sounds like a human extremophile.

Extremophiles, from the Latin *extremus* meaning "extreme" and the Greek *philiā* meaning "love," are organisms that thrive in extreme conditions that would be detrimental to most life on Earth. One of the best examples of an extremophile is the tardigrade, or water bear. If you Google "tardigrade," don't let their appearance fool you. Tardigrades are pretty indestructible. True, they have couch-like bodies with four pairs of stubby, poorly articulated legs and only range in size from 0.012 to 0.020 inches (though the largest species may reach a heady 0.047 inches), but check this out: scholars have found them on top of Mount Everest, in hot springs, under layers of solid ice, and in ocean sediments. They're able to survive the most extreme environments that would kill almost every other animal. They've survived temperatures as hot as 303°F and as cold as -457°F. They can go without water for ten years, they can stand one thousand times more radiation than other animals, and they've even been known to survive the vacuum of space. For ten days in 2007, tardigrades were taken aboard the Foton-M3 mission into low-Earth orbit where they were exposed to the hard vacuum of outer space. Did that phase them? Not in the slightest. On their return to Earth and after re-hydration, most of the tardigrades recovered in just thirty minutes. They have been immersed in chemicals and squeezed by pressures six times greater than those at the bottom of the ocean; but, like living granules of instant coffee, with one drop of water they come back to life—even a century later.

With creatures like tardigrades about, it's little wonder that scholars take the science of super-healing seriously. Consider the question of Deadpool's cancer and regenerative power. Scholars agree that cancer cells grow very swiftly indeed. After all, it's one of the reasons the disease is so deadly. Malignant cells draw other cells into giant tumors and prevent the body from performing its usual functions. The secret to this spread of cells lies with genes known as oncogenes, which multiply and disperse all over the body. And yet, if things are working normally, tumor-suppressor genes stop proto-oncogenes or normal genes from going haywire. So perhaps the secret to Deadpool's regeneration power is this: When he gets a limb lopped off or takes a bullet in the butt, oncogene expression mushrooms in that area, with the cancer cells multiplying and renovating the limb or tissue. Later, when the healing process is almost done, tumor-suppressor genes activate and start to stifle the oncogenes before they get too out of hand.

Nature has managed to conjure an animal that uses this relationship between oncogenes and tumor-suppressor genes to regenerate. The axolotl, also known as the Mexican salamander, is able to regrow limbs, sections of spinal cord, and even segments of its brain. In fact, you can cut its spinal cord, squash it, and take away a section, but it will *still* regenerate. It's the same story with the salamander's limbs; they can be severed at any length—wrist, elbow, upper limb—and regenerate a perfect replacement with no omissions or skin scarring at the site of amputation with every tissue replaced. Not only that, but they're able to regenerate the same limb as many as one hundred times, perfect every time. (Yes, I had the same thought: What kind of scientist repeatedly lops the limb off the same sad salamander? Poor little fella.)

Scholars are studying the axolotl in the hope of applying its incredible salamander-healing powers to humans someday. Modern molecular tools and techniques, which enable technicians to define the cells and genes used in salamander regeneration, will help shed light on the future treatment of injuries and diseases in humans. After all, some time ago in the very distant and ancient misty past, humans and salamanders shared a common ancestor, but scholars face the possible limitation that the subsequent evolution of both species has been too divergent to get a salamander fix on future human healing. So, though the mission may not be to make Deadpools of us all, there's great healing potential in trying to harness axolotl-power in humans.

DAILY DIARY: SUPER STRONG! LIVING WITH HULK'S INCREDIBLE STRENGTH

"What is strength, without a double share of wisdom? Vast, unwieldy, burdensome; proudly secure, yet liable to fall by weakest subtleties; not made to rule, but to subserve where wisdom bears command."

—John Milton, *Samson Agonistes* (1671)

"Strength does not come from physical capacity. It comes from an indomitable will."

—Mahatma Gandhi

"Dr. David Banner, physician/scientist, searching for a way to tap into the hidden strengths that all humans have. Then an accidental overdose of gamma radiation interacts with his unique body chemistry. And now, when David Banner grows angry or outraged, a startling metamorphosis occurs."

—Narrator, *The Incredible Hulk* (1978–1982)

He once destroyed an asteroid twice the size of the Earth. He is said to have held aloft a 150-billion-ton mountain, and when on the planet Sakaar, he once dragged two tectonic plates together and by sheer strength alone stopped the planet from splitting apart. Hulk's strength is the stuff of legend. Said to possess a superhuman strength of the highest level, the Hulk derives his power when subjected to emotional stress, and his level of strength is often portrayed as proportionate to his level of anger.

But what's the real daily story of being super strong? Just imagine living with Hulk's incredible strength. Say you woke up one morning with Hulk power, more than one thousand times stronger than the day before. We can forget problems of provenance. When Marvel first created the Hulk character in May 1962, the first issue of *The Incredible Hulk* fused elements of Mary Shelley's *Frankenstein* with the tale of Ben Grimm, the Thing from *The Fantastic Four*.

Researchers at Columbia University recently found the gonorrhea bacterium is, pound for pound, the strongest living creature on Earth. Able to pull up to one hundred thousand times its own body weight, gonorrhea is hugely stronger than any other creature ever recorded. (The oribatid mite, relegated to a poor second behind gonorrhea, can only lift around one thousand times its body weight, which makes it many times weaker.) And check out gonorrhea power: a heavy horse with this strength could pull one hundred million kilograms of weight without the aid of wheels; a 168-pound human packing this kind of power could comfortably pull the Eiffel Tower behind them!

Let's say you've got gonorrhea power. But the origin of your newfound strength isn't our concern here—*living* with it is. For that matter, we might as well go the whole nine yards: Let's imagine you wake up six feet six inches tall and weighing around one thousand pounds (we might leave the body color to the side, as that can result in problems with personas). Now, how exactly would you handle those delicate daily tasks? Don't you think you might soon neglect notions of nuance?

The world would be so different once armed with Hulk strength. Suddenly, all things seem so fragile. Your strength has scaled up more than one thousand times. You'd need to take care of high-fiving your friends in case you break their fingers. And be lenient on lovers too, as you could break their bones from even the gentlest of hugs. Dining out would be difficult; you might easily drive that chopstick through the sweet and sour pork *and* the plate. Out of the blue, table manners become as tedious as neurosurgery.

THE DAMSEL-IN-DISTRESS CLICHÉ

And don't go rescuing damsels in distress, hero boy. Sure, she may be falling from a skyscraper, thrown out of the penthouse by yet another supervillain out to rule the world by fair means or foul, but catching her has its drawbacks. Within seconds you'll be cradling her lifeless form in your arms. Why? Physics, bro.

Say, for no apparent reason, you tried hammering a bowling pin into your bedroom wall. Come to think of it, with your Hulk strength, you probably don't need that hammer. Pummel that pin! But the pin won't get half as far as the nail you hammer in next (naturally, the bowling pin is ruined by your power, and your neighbors desperately want to know what's going on). But why is the nail easier to drive home than the pin, using the same force? Pressure. When you pummeled the pin with your palm, the force was spread over the entire area of the bottom of the bowling pin. Outcome: the pin doesn't get very far, but surely shatters on impact. But, when you use the same force on a nail, the nail is far more likely to pierce the wall because the magnitude of the force is now acting through a smaller area—the point of the nail. In a nutshell, pressure is force divided by area. So, the smaller the area, the bigger the pressure. It's the same reason we can lift heavy objects without piercing our skin, but a judicially placed hypodermic needle can bleed us with just a little prick.

Let's go back to that damsel in distress. You can now see why the pressure exerted on her body can be worked out by dividing the force of her impact by the area of the top of your arms with which she comes in contact when you catch her. Your newfound strength doesn't come into play. Your arms are strong enough to catch her, but not without breaking her bones. Her spine is simply not strong enough to be caught by you without being snapped.

Plan B: With the super speed you probably don't possess, you could swiftly rip off the nearest door. Then, thinking ahead (and all this before she drops), you use the door to make a bigger area with which to catch her. But plan B fails too, as it's not the fall that kills her, but the sudden stop at the bottom.

DAMSEL DROP

Consider the damsel drop in detail. Say she's thrown from a forty-story building, which is about 375 feet. With your fresh Hulk dimensions, you're now six feet six inches tall, but on tiptoe and maybe with a little Hulk leap with your arms above your head and holding the door, you could possibly make a height of fifteen feet. But your hopes of spreading the pressure over the larger surface area of the door are fated to failure as all you're really doing is moving the ground up by fifteen feet.

So, the damsel drop is now 360 feet rather than 375. If you don't count the drag on the damsel due to air resistance, then she'll reach a terminal speed of over 150 feet per second just before impact. That's the same as crashing at over one hundred miles an hour into a wall with a door conveniently placed in your way.

Even if you add super flight to your powers, there's still physics to conjure with. Yes, you could rescue the damsel drop with a swift ascent, but here's what you'd have to do. First, you'd approach her altitude by flying up to where she is, then you'd have to start flying down at the same speed with which she is falling. Lastly (and hopefully not fatally!), you'd grab her and gradually decelerate until you hit the ground together.

Naturally, our solution comes with some crucial clauses. You'd have to ensure plenty of cushion space between the point the damsel starts her drop and the ground. That might not be easy. And don't go wasting time slinking into your superhero flying costume. Every second you waste before flying up to her height, her pretty damsel cranium is getting closer to the ground. Realistically, since the damsel is dropping from up on high, and you can't get to her until she's quite close to the sidewalk, there's really little you can do other than watch her damsel delights turn magically into mulch. It seems old Milton was right about super strength—to be taken only with a double dose of super wisdom!

DAREDEVIL: HOW FAR ARE THE OTHER SENSES HEIGHTENED BY BLINDNESS?

KAREN PAGE: "How did it happen?"

MATT MURDOCK: "Car accident. When I was nine."

KAREN PAGE: "Must have been rough."

MATT MURDOCK: "No. I made it through."

KAREN PAGE: "Do you remember what it was like . . . to . . . to see?"

MATT MURDOCK: "I, um. . . . Yes, I remember."

KAREN PAGE: "I can't imagine what that must be like."

MATT MURDOCK: "You know, I'm supposed to say I don't miss it. That's what they teach you in trauma recovery. Define yourself by what you have, value the differences, make no apologies for what you lack. And that's all true for the most part . . . but it doesn't change the fact that I . . . I'd give anything to see the sky one more time."
—*Into the Ring* written by Drew Goddard, *Daredevil 1.01* (2015)

CLAIRE TEMPLE: "How do you . . . I mean, I know that you're blind, but you see so much. How?"

MATT MURDOCK: "I guess you have to think of it as more than just five senses. I can't see, not like everyone else, but I can feel. Things like balance and direction. Micro-changes in air density, vibrations, blankets of temperature variations. Mix all that with what I hear, subtle smells. All of the fragments form a sort of impressionistic painting."

CLAIRE TEMPLE: "Okay, but what does that look like? Like, what do you actually see?"

MATT MURDOCK: [pause] "A world on fire."

CLAIRE TEMPLE: "If all I saw was fire, I'd probably want to hit people too."

—*World on Fire* written by Luke Kalteux, *Daredevil 1.05* (2015)

"Every surface has its own acoustic signature— I can recognize a tree, for example, because the trunk produces a different echo from the leaves. The hard wood reflects the sound, whereas the leaves reflect and refract, too, scattering the sound waves. Everything around me becomes identifiable with a click. It provides me with a 3D image in my mind with depth, character, and richness; it brings light into darkness. I can often find my way out of an auditorium quicker than a sighted person because I can identify the exit. If I'm in a noisy place such as a concert, I don't feel anxious—I just increase the volume and my click cuts through the noise. I'm very familiar with its sound and don't feel at all self-conscious if other people hear me. I don't have superhuman hearing, even though I'm sometimes called Batman; I have just trained my ears to understand the echoes. Anyone could do it, sighted or blind—it's not rocket science."

—Daniel Kish, "I taught myself to see," *The Guardian,* 13 July 2013

Hell's Kitchen has thrown up some notable residents over the years: James Cagney, James Gunn, Robert Fripp, Larry David, Sylvester Stallone, Mickey Rourke, Mario Puzo, Stanley Kramer, Frank Miller, and Alicia Keys, to name just a few. And yet, for many people, Hell's Kitchen's most famous resident is Matt Murdock and his superhero alter ego, Daredevil. Born in this bastion of poor, working-class Irish Americans, Murdock was abandoned by his mother and raised by his boxer father, "Battling Jack" Murdock. Being raised in an area with an infamously gritty reputation led Murdock to realize that the rule of law was needed to prevent people from behaving badly. But young Matt's decision to study law was punctuated by a terrible accident that left him blind. In a valiant effort to save a man from a looming lorry, a toxic and radioactive cargo was spilled, which left Murdock without his sight and reliant on his remaining senses. Under the strict instruction of a blind martial arts master in Stick, Murdock mastered his heightened senses and transformed himself into a formidable fighter.

Now, many superheroes have it easy. Superman simply moves to a sunnier environment and, boom, he's faster than a speeding bullet, more powerful than a locomotive, able to leap tall buildings in a single bound, etcetera, etcetera. Others are lucky enough to be born with mutations that bless them with powers. Storm came from a long line of African witch-priestess mutants who were born with the superhuman ability to control the weather. But Matt Murdock had to work at his skills, and work hard. It's one thing combatting evildoers if you have the power of flight, super strength, or weather wielding, but what if you can't even boast a full set of functioning eyes? Does it become a surefire situation of "there goes the neighborhood?"

THE EMPIRE OF THE AIR

When Murdock is Daredevil, an otherwise ordinary human becomes a kind of flawless receiver, a being able to sense and gather all available data, much of which is usually invisible to us. It may help to think of Daredevil as a kind of radio receiver, scanning all wavebands of the world in search of a signal in all that chaotic static. And when he does detect something in his signal-to-noise ratio, he hones in on his oral quarry, tuning into specific targets such as a subtle sound or the scent of an open wound. But Daredevil is no ordinary radio receiver—he can tune into twenty "wavebands" at the same time, and paint a mental picture of the haunt in which he finds himself.

Tutorial 101 for the young Murdock under the tutelage of Stick must surely have been echolocation. Before any fearless crime-fighting can be done, Murdock would first have learned simply how not to bump into things. Echolocation is basically the ability to "see" with sound; it's biologically based sonar (Sound Navigation and Ranging).

Imagine Murdock is on vacation at the Grand Canyon, away from the bustling streets of Hell's Kitchen. If he were to shout across the canyon, his vibrating vocal chords would produce a sound wave, which would be echoed back to him off the face of the rocks on the other side of the canyon. If air pressure and composition are normal and constant, sound waves move at the same speed. So, Murdock would have learned to judge this speed of sound, and use the sound and his innate sense of time to gauge the distance across the canyon.

Echolocation is the technique used by animals such as dolphins, and especially bats. In fact, studying bat vision gives a good idea of the way in which Daredevil would use the basic principle of echolocation. Bats make sounds the same way humans do; they run air over their vibrating vocal chords. Bats either make sounds through their mouth, which they droop open as they fly, or through their nose. Scholars think that some bats' strange nose structure helps focus the sound for more precise pin-pointing of their target prey.

With most bats, their echolocation sound is so high pitched that it's outside the range of ordinary human hearing. But, like Matt Murdock's shout across that canyon, bat sound still behaves and travels through the air as a wave energy that bounces off objects and surfaces. A bat sends out a sound wave, and then listens intently to the echoes it returns. By natural means, the bat's brain is able to marshal the returning data and work out how far away an object is.

Like Daredevil, the bat can also work out the location of an object, it's size, and even the direction in which it's going. It knows if an insect is to its right or left by deciphering the sounds coming to its right ear compared to the sound reaching its left ear. For example, if the bat picks up an echoed sound at its right ear before its left, then the insect is clearly to the right. To help with such deciphering, bat ears have an advantage that Daredevil lacks—their ears contain a complicated collection of folds that aid its determining an object's vertical position.

How do bats measure the size of their prey? A bat can gauge the size of an insect when it processes the echo data from an object. Here, *intensity* of the echo is key. A smaller object reflects back less of the sent wave and so will create a less intense echo. And how do bats sense the direction of their prey? Here, the *pitch* of the echo is crucial. When an insect is moving away from the bat, the echo will have a lower pitch than the sent sound. When an insect is moving toward the bat, the echo will have a higher pitch. And so, the bat's kingdom is the very atmosphere in which it hunts, and creatures that use echolocation live less in a world on fire and more in an empire of air.

A WORLD ON FIRE

The bat intuitively channels all its data in the same automatic way humans manage the audio-visual data gathered with our ears and eyes. Scholars believe the bat conjures an echolocation image in its brain which is akin to the images humans form in their heads based on sight. But to what extent does the case of the bat paint a perfect picture of what Daredevil is meant to be doing? The answer is, not entirely.

The basis for Matt Murdock's powers is more than just echolocation. Rather, Daredevil's skills derive from an ability to focus on more than just radar-like data. He can scent and feel more of the world than normal humans, too, and it's his synthesis of all of these senses that give him his crime-fighting gifts. Sure, echolocation may come first, with its ability to pinpoint enemies in space, but then comes the panoply of other powers to make that mental map of a world on fire.

Why not try your own Daredevil experiment at home? Try making your way around your living space with your eyes closed. Even if you end up with a headache, before you get to that point, think about what you experience as you move (slowly) about your house. When you close in on another object, you get that rather unnerving tingle, which is the body's warning that you are about to bump into something. The phenomenon has sometimes been called "facial vision" by blind patients in the past, naming the experience after feeling a kind of pressure on their face when something is nearby. Scholars believe these blind patients were interpreting sensory data and processing it as tactile pressure and feeling that they're about to bump into something. It's a tantalizing glimpse of the way in which humans might be able to become like Daredevil.

When thinking about the powers that Daredevil has, it's worth remembering that most humans have more than just the famous five senses of taste, sight, touch, smell, and sound. Let's try another experiment. Close your eyes and hold your hand in front of your face. You can kind of feel that the hand is right in front of you. That sense is known as proprioception, and it's one of the senses beyond the more famous five. And these latent senses are exactly the kind of skills that Daredevil had to learn to harness as part of his streaming of ambient data.

To get a good idea of a real-life Daredevil, consider the case of Daniel Kish, an American expert in echolocation and president of World Access for the Blind (WAFTB). Kish is blind, and yet freely rides a bicycle through his neighborhood. Kish teaches others to echolocate. He explains in an article from *The Guardian*, "If you hold up a book in front of you and click, then take it away and click, you can hear a difference, just as you know you're in an empty room because it's echoey. When I was in college I wrote my thesis on echolocation, and during my research I had to consciously deconstruct how I was doing it to understand the process. I know there's a wall in front of me, I'd think, but what's tipping me off? I would set myself tasks and try to get quicker and quicker at navigating obstacle courses." When scholars have performed brain scans on people like Daniel Kish, they've found that, when they're making the clicks and listening for the echoes, areas of the brain related to visual and spatial processing are lighting up too, and not just auditory areas of the brain. As Kish says, "I have made it my life's work to teach blind children how to empower themselves using echolocation, which I call flashsonar. As you become more adept, you also click more subtly and naturally, like blinking, so often people around you aren't aware you're doing it and you aren't stigmatized for it. Now I can ride along a busy street or go on a trail in the woods. I have never hit a pedestrian—touch wood—because I don't ride on the pavement. Cars are excellent echo targets, so I can easily avoid them. I won't say I've never had an accident, but every activity holds an element of risk. Negotiating rush hour traffic isn't my dream; I am just glad I can if I want to."

So, scholarly research suggests the human brain is plastic enough, it can transform to re-navigate data and make alternate sense of the same data. Our brains can bring the blind closer to a Daredevil than most people imagined. Okay, the ninja fighting skills would need a lot more work, but in a very real sense all humans have the potential to be like Daredevil. All the brain needs is a little bit of re-wiring, and it can transform and do the kind of incredible things associated with some version of a superhero. As Daniel Kish says, "Anyone could do it, sighted or blind—it's not rocket science."

HOW COULD LUKE CAGE'S SKIN BE IMPENETRABLE?

"A wrongfully charged inmate in Seagate Prison named Carl Lucas undergoes a scientific experiment. When a racist prison guard tampers with the machine, Lucas develops some supernatural side-effects! Namely, enhanced strength and bulletproof skin! Luke Cage is born."

—*Hero for Hire* written by Archie Goodwin, *Luke Cage #1* (1972)

LUKE CAGE: "Man deserves his parole just for walking in here, doc. Looks like strictly mad scientist territory—includin' a bathtub for Frankenstein's monster!"

DOCTOR NOAH BURNSTEIN: "The equipment was made by Stark Industries, Lucas—as part of a research grant given me. It's an electro-chemical system for stimulating human cell regeneration."

—*Hero for Hire* written by Archie Goodwin, *Luke Cage #1* (1972)

LUKE CAGE: "You even know who Crispus Attucks was? A free black man. The first man to die for what became America. He could've acted scared when those Brits raised their guns. Blended in, in the crowd. But he stepped up! He paid with his life. But he started something. That's what Pop did. Not me. I laid in the cut until he stepped up! And it cost him his life, too. I ain't laying back no more! You wanna shoot me? Do it."

—*Code of the Streets* written by Cheo Hodari Coker, *Luke Cage 1.02* (2016)

LUKE CAGE: "This burden is bigger than you. Or me. People are scared but they can't be paralyzed by that fear. You have to fight for what's right every single day, bulletproof skin or not. You can't just not snitch, or turn away, or take money under the table because life has turned you sour. When did people stop caring? Harlem is supposed to represent our hopes and dreams. It's the pinnacle of black art, politics, innovation. It's supposed to be a shining light to the world. It's our responsibility to push forward, so that the next generation will be further along than us."

—*You Know My Steez* written by Aïda Mashaka Croal & Cheo Hodari Coker, *Luke Cage 1.13* (2016)

"Body organs aren't all internal like the brain or the heart. There's one we wear on the outside. Skin is our largest organ—adults carry some 8 pounds and 22 square feet of it. This fleshy covering does a lot more than make us look presentable. In fact, without it, we'd literally evaporate. Skin acts as a waterproof, insulating shield, guarding the body against extremes of temperature, damaging sunlight, and harmful chemicals."

—"Skin," *National Geographic* online

It's just as well Luke Cage is bulletproof. America's gun culture is the most dangerous in the world. According to CNN, Americans own nearly half (48 percent) of the estimated 650 million civilian-owned guns worldwide. Americans also own more guns per capita than residents of any other country. Even though the US makes up less than 5 percent of the world's population, it can still boast 31 percent of global mass shooters. Gun homicide rates are more than twenty-five times higher in the US than in other high-income countries. And even if Luke decided to take a vacation south, he'd find America's neighbors suffering from a similar contagion: worldwide, the countries with the highest gun-homicide rates are in Central and South America.

Luke Cage was the first black superhero to be featured as the protagonist and eponymous hero of a comic book. With an alter ego of Carl Lucas, Luke Cage was created by Marvel Comics in 1972 during the height of the Blaxploitation movement, an ethnic subgenre of the exploitation movie, initially made for an urban black audience, but whose appeal soon broadened beyond racial and ethnic borders. Luke's powers and abilities were drawn from science. Even before the lab experiments, Luke Cage was an exceptional street fighter and gifted athlete. His superhuman abilities came about as a result of a cellular regeneration experiment, which greatly fortified his body. He suddenly possessed superhuman strength and stamina and became highly resistant to physical injury. In particular,

Luke's skin became resistant to bullets, puncture wounds, chemical and biological attacks, and extreme temperatures and pressures without any lasting injury.

Luke's impenetrability is usually put down to his extremely dense skin and muscle tissue. But that can't be the source of Luke's powers. Adding more mass to the same volume of Luke's skin to increase its density wouldn't do anything for his powers. Density is mass divided by volume, so the only way to increase Luke's density (since we don't really want to shrink his volume down to a tiny "ant-man" Luke, given those ever-violent streets) is to increase his mass. And doing that would merely add to his skin's weight and, if anything, simply make a more lumbering Luke, now ponderously prowling the streets of Harlem in a hoody.

LUKE THE SPACEMAN

In the Netflix version of *Luke Cage*, our superhero has both incredible strength and impenetrable skin. He has survived a shotgun blast to the face and artillery of various types and power. So, with all this gunfire cracking off about his body, how strong would Luke's skin have to be to keep bullets at bay, and what scientific secret might make his skin that impenetrable?

Now, human skin comes in three layers: epidermis, dermis, and subcutaneous. The outermost layer, the epidermis, is a waterproof barrier that also creates our skin tone. The dermis, beneath the epidermis, contains the tough connective tissue, hair follicles, and sweat glands. The deeper subcutaneous tissue, or hypodermis, is made of fat and connective tissue. In Luke Cage's case, how far into his skin is the bullet probing?

A bullet gains entry into the body by applying a concentrated pressure to the skin at a particular point. As the pressure applied is the force of the bullet acting over that certain area, Luke's skin would simply have to be tough enough to resist the force of the bullet that is trying to push through, so our calculations need to work out just how tough is enough for impenetrability.

Consider one of the world's most commonly used bullets, the nine-millimeter. This nine-millimeter diameter bullet comes in many varieties, but on average it has a mass of about eight grams and, after it's been shot out of a gun, has a speed on impact of around 350 meters a second. Luke's bulletproof skin must be strong enough so that the nine-millimeter bullet comes to a stop somewhere in his skin, let's say a hair's breadth in, before it makes it all the way through to cause havoc to his internal organs.

Calculations based on this average data of a nine-millimeter bullet traveling at 350 meters a second and stopping after just a hair's breadth of penetration suggest that Luke's skin strength is able to fend off a force of 2 billion Newtons per square meter. Now you may be wondering what on earth the force of a Newton feels like; a commonly used measure is that a Newton is about the weight of an apple on Earth. It could be that this myth came about as people like to tell the tale of Newton's apple and his realization of gravity, but if you weigh some sample apples, no matter how petite an apple you pick, each apple weighs in at around two Newtons apiece. So, Luke Cage's skin is tough enough to withstand the weight of a billion apples per square meter of his skin!

All Luke need do to become impenetrable is make like a spaceman.

And here's why: Even though fending off 2 billion Newtons of force sounds like the most troublesome of tasks, there is a material that possesses what scientists call the ultimate tensile strength, which is able to resist this awesome amount of pressure. This material has an ultimate tensile strength above spider silk, even above diamond, and it's called Kevlar. In Daily Diary: The Need for Speed—How the Flash Deals with Dynamics (pg. 47), we talk about the speed of micrometeorite debris in Earth orbit, the kind of debris hurtling toward Sandra Bullock in the 2013 movie *Gravity*. Those micrometeorites are traveling "faster than a speeding bullet," and are the kind of cosmic hazard from which orbiting astronauts need protection, as they speed along at 5 miles per second. The material of choice that keeps those astronauts safe is Kevlar. When astronauts launch off Earth from the blue into the black there's some serious dressing up to be done. Today's spacesuit, worn by all fashionable space explorers, is known as an EMU (Extravehicular Mobility Unit), is thirteen layers thick, and costs a cool 12 million dollars each. And the key component of the spacesuit's eight-layer thermal micrometeoroid garment is Kevlar, the material also used in bulletproof vests.

What gives Kevlar its ultimate tensile strength? The short answer is carbon. Rather appropriately for an element that is to be found in spacesuits, carbon is cosmic. Not only does it form the basis of life on Earth, easily bonding with life's other main elements, but carbon is also light and small, making it an ideal element for creating the longer and more complex chemicals of life such as proteins and DNA.

Carbon's chemistry has the ability to form long chains of molecules called polymers. Kevlar is a long chain-like polymer that consists of repeating units called monomers. A Kevlar fiber is an array of molecules organized parallel to each other, somewhat like a bundle of uncooked spaghetti. This system of untangled molecules is called a crystalline structure, and the crystallinity of the Kevlar polymer strands contributes significantly to Kevlar's unique strength and rigidity, helping make Kevlar bulletproof.

Someone should have taken the time to weave a little Kevlar into Luke Cage's backstory. Perhaps during the experiments conducted by Noah Burnstein or some other mad doctor, a polymer like Kevlar was somehow synthesized into Luke's skin layers, gifting him that superhero status in Harlem and beyond.

DAILY DIARY: HOW DOES THE HULK DEAL WITH BODY MASS?

"Atoms are very small—one hundred million of them end to end would be as large as the tip of your little finger. But the nucleus is a hundred thousand times smaller still, which is part of the reason it took so long to be discovered. Nevertheless, most of the mass of an atom is in its nucleus; the electrons are by comparison just clouds of moving fluff. Atoms are mainly empty space. Matter is composed chiefly of nothing."

—Carl Sagan, *Cosmos* (1980)

"I am a collection of water, calcium, and organic molecules called Carl Sagan. You are a collection of almost identical molecules with a different collective label. But is that all? Is there nothing in here but molecules? Some people find this idea somehow demeaning to human dignity. For myself, I find it elevating that our universe permits the evolution of molecular machines as intricate and subtle as we."

—Carl Sagan, *Cosmos* (1980)

"A physicist is the atom's way of thinking about atoms."

—George Wald, *Life and Mind in the Universe* (1984)

TRANSFORMATION 1

Bruce Banner stumbles down a corridor, his eyes turning emerald green. He screams as the transformation begins. His watch flies off his wrist, pant pockets rip, his feet mushroom and tear off his shoes. As he stomps and staggers forward, his growing torso now so huge that his shirt is left in tatters, he becomes the Hulk . . .

TRANSFORMATION 2

Banner's head begins to turn green and widen. His entire form burgeons in size: neck and hands, abs and biceps, legs and feet. He slams his fists onto the floor, and the entire room quakes. He is forced to leave the house, as he continues to morph . . .

TRANSFORMATION 3

Banner wakes up from a nightmare and finds himself inside a water tank. His transformation triggered, his skin turns green, his muscles grow, and the straps on his legs, body, and arms rip apart as he becomes so big he bursts out of the tank.

Marvel had always gone for gigantism. Back in the days of American and Japanese monster movies, Marvel was known as the print medium of the giant monster. Creatures such as *X: The Thing That Lived*, *It: The Living Colossus*, and *Gor-Kill the Living Demon* clomped over the monster-packed pages of Marvel Comics. Then, in late 1961 and early 1962, the monsters were replaced by superheroes. But even now, Marvel hasn't dropped the monster. With the Incredible Hulk and the Fantastic Four's Thing, Marvel merely mashed up two genres into one.

"The strangest man of all time!" "Is he man, or monster, or is he both?" "Fantasy, as you like it!" were the captions on the cover of the very first Incredible Hulk. The new trend was for men to take on monster proportions. Fast-forward to the CGI'd twenty-first century and we see that, in movies such as Marvel's 2012 blockbuster *The Avengers*, Bruce Banner is still morphing into the monstrous Hulk. But, as a first-rate scientist like Dr. Banner would ask, where's all that extra mass coming from? What's the source of the material, which makes good his monstrous transformation?

BANNER BASICS

Let's consider the Banner basics of becoming the Hulk.

In physics, there's a Law of the Conservation of Mass that says that mass can neither be created nor destroyed. Take any isolated system defined by a boundary that matter and energy cannot cross. The system could be a bunch of water, the entire cosmos, or Bruce Banner himself. But, inside each system boundary, mass (which can also be considered as matter or energy) can neither be created nor destroyed.

Water is a good basic example. Water is made of millions of hydrogen and oxygen atoms in a ratio of two to one respectively. Let's say we have ten grams of this water. Now, whether the water boils away to steam or freezes into ice, it will remain just ten grams—ten grams of steam, or ten grams of ice. In short, the mass of particles that makes up the water stays the same. Only the state of the particles changes (water into steam, or water into ice), which merely means their closeness, arrangement, or motion has changed.

Bruce Banner, to the best of our knowledge, is an isolated system. And the Law of the Conservation of Mass dictates that Bruce's mass should not change just because his size changes as he morphs into the Hulk. It seems sacrilegious to compare the Hulk to cupcake—but let's do it anyway. A decent baker, using the *Marvel Superheroes' Cookbook*, would know that, when they bake a fluffy cupcake, even though their delicious bake-off is much bigger than the cake mix that went into the oven; the mass of the cake mix will still equal the mass of the cake, plus a little moisture that evaporated during baking.

Scale up from cupcake to Hulk and the same rules apply. Let's say that in *The Avengers*, Dr. Banner is around six feet tall, and the Hulk around nine feet. Don't worry about the exact dimensions. What's important here is the chemical equation of Banner > Hulk. In chemistry, molecules react to make new structures, but all the chemicals should be accounted for. The scale up we need is a factor of 1.5x. That's the difference between the 6 feet of Banner and the 9 feet of Hulk (6 feet x 1.5 = 9 feet).

SQUARING UP TO THE MASSIVE

Next, we invoke the Square Cube Law. This math idea accounts for the connection between the area and volume of an object as that object increases or decreases. The Law is usually attributed to the famous Italian physicist, Galileo Galilei, and dates back at least to the year 1638. The Law says that as a body grows in size its volume grows even faster. The Law has lots of applications, including the explanation of why elephants have a harder time cooling down than mice, and why the world's engineers find building taller and taller skyscrapers progressively more difficult.

The Square Cube Law also applies to the Hulk. When Banner expands from 6 feet to 9 feet tall, Galileo's Square Cube Law shows us that his volume will be 3.375 (1.5 x 1.5 x 1.5 = 3.375) times his normal volume, since Banner expands in 3D when he morphs into the Hulk. So, what happens to Banner's body mass? Well, a person's mass is equal to their density, multiplied by their volume. And since Banner's volume has increased by 3.375 times, that means his mass must have also increased in size, assuming his density stays the same. But, to be precise, there are two possibilities when Banner superheroes into the giant Hulk, depending on density. Possibility #1 is that Banner's density decreases and his body mass stays the same, and when Banner expands into the Hulk, his bony and fleshy human density morphs into something akin to an air-boned Barney the dinosaur. Possibility #2 is that Banner's mass increases by the same 1.5 times. And that means Banner's weight as a human, say 200 pounds, becomes 3.375 times greater when superhuman, which is 675 pounds, which is more than a third of a ton! In that case, Banner's bones would be in danger of breaking, not bending, and his tendons would run the real risk of tearing.

Possibility #2 seems far more plausible, so no wonder paving stones crack under the weight of the Hulk's hulk. No wonder he ruins lawns in every neighborhood he wrecks, and no wonder he has to buy those super pants, which always seem flexible enough to stay wrapped around his ass after expansion. And, assuming the Hulk still functions like a mammal, his heart will have to work harder, too, as it needs to pump enough oxygenated blood around a body 3.375 times bigger than it was. That means more energy from calories. Check the math: An average intake of calories for a very active human male is around 3,000, but scale up once more by our factor of 3.375 and the daily calorie rate becomes a whopping 10,000! That's a green light right there for around 18.5 Big Macs a day, thirty-six Snickers, or 67.5 servings of Doritos Cool Ranch Tortilla Chips. Our hero might spend so much time stuffing his face, he'd have little time left to fight crime. Luckily, though, the transformation is temporary, but it just goes to show there's more to Hulk morphing than just height.

So much for flesh and bone. What about heroes who morph their bodies into rock? Well, as we said at the start of this subject, all things in the cosmos are composed of atoms. And atoms are arranged into elements. The famous Periodic Table of chemistry is organized into elements, and each successive element differs from the one before in terms of the number of protons in the nucleus. Hydrogen comes first with just one proton, then helium with two protons, lithium with three protons, and so on. These three elements are the only ones made in the original Big Bang, or so the story goes.

But what about a human doing a big bang into a superhero made of sand or rock? The main component of the most common sand is silicon dioxide. And many rocks are silicates, also composed of silicon and oxygen. Meanwhile, our bodies are made up of 65 percent oxygen, 18 percent carbon, 10 percent hydrogen, and 7 percent of many other elements. Only a trifling 0.002 percent of the human body is silicon. Since chemical reactions are simply recombinations of constituent elements, where is all the extra silicon actually coming from to provide the rock and sand that makes our superhero complete?

Nuclear fusion will do it, but nuclear fusion's not so much a *chemical* reaction, which usually involves electrons, as it is a *physics* kind of equation, which alters the elements by changing the very nucleus at the heart of the atoms. And that needs so much heat, the only normal and natural occurrence of this fusion is the very core of stars. So, to get his atoms fused in a moment, the core temperature of our superhero would need to be hotter than the Sun. Onlookers would get blasted with a Nagasaki, and our hero would essentially be a mobile nuclear furnace, radiating every damsel in distress and vaporizing every person he tried to save.

So, you see, body mass must be figured in to any supersizing to superhero status.

WOULD HUMAN ENHANCEMENT CREATE SUPERMEN OR SUPER TYRANTS?

"We spend a great deal of time studying history, which, let's face it, is mostly the history of stupidity. So, it's a welcome change that people are studying instead the future of intelligence. The potential benefits of creating intelligence are huge. We cannot predict what we might achieve when our own minds are amplified by AI. Perhaps with the tools of this new technological revolution, we will be able to undo some of the damage done to the natural world by the last one—industrialization. And surely, we will aim to finally eradicate disease and poverty. Every aspect of our lives will be transformed. In short, success in creating AI could be the biggest event in the history of our civilization. But the creation of powerful artificial intelligence will be either the best, or the worst thing, ever to happen to humanity."

—Stephen Hawking, *The Guardian* (2016)

"I'm increasingly inclined to think that there should be some regulatory oversight, maybe at the national and international level, just to make sure that we don't do something very foolish. I mean, with artificial intelligence, we're summoning the demon."

—Elon Musk, *MIT's AeroAstro Centennial Symposium* (2014)

"I don't want to really scare you, but it was alarming how many people I talked to who are highly placed people in AI who have retreats that are sort of 'bug out' houses, to which they could flee if it all hits the fan."

—James Barrat, *Our Final Invention: Artificial Intelligence and the End of the Human Era (2013)*

"The truth of the world is that it is chaotic. The truth is, that it is not the Jewish banking conspiracy or the grey aliens or the 12-foot reptiloids from another dimension that are in control. The truth is more frightening. Nobody is in control. The world is rudderless."

—Alan Moore, *The Mindscape of Alan Moore (2003)*

Supervillains always seem on edge, don't they?

Take Wilson Fisk, for example. There are few villains edgier than Fisk. He has no special powers. Nor has he the kind of global or cosmic reach that sits in the toolkit of other more exotic supervillains. Fisk grew up poor and was bullied in school, but the worm turned when he became the Kingpin of Crime. He lords it over New York City with an iron grip. But Fisk is a deeply troubled individual, and he doesn't hide his mania when he's forced to deal with people directly. On other occasions, he seems grounded and realistic, making Fisk perhaps the most scarily human of supervillains. At the other end of the supervillain scale is Loki, the "God of Edge." As supervillain credentials go, being a Norse god ain't a bad start. Loki, actual God of Mischief and half-brother to Thor, has been around a lot longer than most other villains. His hatred and jealousy of Thor runs so deep that Loki will stop at nothing, even the destruction of Asgard itself. And in the league of master manipulators, Loki is king.

Then there's Victor Von Doom. Old Victor was one of the main inspirations for Darth Vader, the greatest villain of them all. Again, Von Doom has no superpowers, unlike his archenemies, the Fantastic Four. He gets by on cutting-edge wit and cunning and an astonishing intellect. And the role of supervillain isn't just a male matter, either—there's gender spread. Take Catwoman. One of the very few characters to know that Bruce Wayne is Batman, Catwoman is cunning and can play any side to her advantage. She serves herself and no other cause or master. She's unpredictable and never plays within the rules. And she frequently crosses the edge between villain and anti-heroine.

But perhaps the greatest, out-on-the-edge human supervillain ever created is the Joker, especially Heath Ledger's stunning portrayal of the Joker in the 2008 movie, *The Dark Knight*. With his head-rattling hysteria and his clown-masked mania, the joker is the uncrowned clown prince of crime. Armed with a puppet-master body language and freaky facial fits, he brilliantly perverts and electrifies every scene he's in. When he's not on screen, you miss his twitching menace. When he *is* on screen, each scene is wrought with tension, and he shoots from the hip such nihilistic lines that are well on their way to becoming legendary pop-culture quotes: "Why so serious?" "I'm not a monster. I'm just ahead of the curve," and especially, "Do I really look like a guy with a plan? . . . Introduce a little anarchy. Upset the established order and everything becomes chaos. I'm an agent of chaos. Oh, and you know the thing about chaos? It's fair." Now, Shakespeare also came up with the odd killer line or two, including the quote from *Henry VIII*, "We all are men, in our own natures frail, and capable of our flesh; few are angels." (Incidentally, it was during a performance of *Henry VIII* at the Globe Theatre in 1613 that a cannon shot, used for special effects ignited the theatre's thatched roof and beams and burned the original Globe to the ground.)

Given that few of us are true angels, human enhancement would seem a dodgy and dangerous prospect, and yet that prospect will soon be upon us. Superhero fiction is based on the science fiction dream that humans will one day transcend our human limits of body and brain. And now science may cross these physical and intellectual barriers through progress in cybernetics and nanotechnology—wetware, as it's sometimes called. It's likely to become reality this century. But would human enhancement create supermen, or super-tyrants?

THINKING UP SUPERMEN

What does the future hold for Man? What will Man, one day, become? Such questions have busied the fertile brains of science fiction writers, filmmakers, and artists.

As science pushes the envelope of technology, notions that once seemed fantastic now find themselves at the forefront of probability. This evolution will no doubt give rebirth to the long-standing pangs of moral peril, as ethics scraps to keep pace with bleeding-edge tech. This dream of human enhancement or augmentation is known as transhumanism. The most common transhumanist belief is that humans will eventually transform into different beings, with abilities so greatly expanded from the natural human condition as to merit the label of posthuman beings. The belief has been explored in philosophy, film, or fiction ever since Darwin.

As we mention elsewhere in this book, German thinker Friedrich Nietzsche had floated the notion of the Übermensch (super-man, over-man, or superhuman) in his 1883 book *Thus Spoke Zarathustra*. Nietzsche's idea is fundamental to the superhero genre, so it'll pop up on occasion. And Nietzsche's notion of the Übermensch was of a being seeking to move "over" its state of being to a greater "stature." No other symbol in science fiction has evolved as dramatically as the super-man. From the most infantile form of human wish fulfillment to more sophisticated anti-hero, the superhero has become an ingenious metaphor of our aspirations and fears for future science.

In the same way that Darwin led science fiction to the alien, theories of evolution have given writers a fantastic framework for stories of super humans. But there is a difference, of course, between Darwin-induced narratives of "fitter" humans and Lamarckian-inspired superheroes whose creative evolution is often instantaneous, and whose newfound powers may well be passed on to their offspring—if they ever had sex.

Creators of these fantastic supermen had originally been surprisingly shy to make their heroes outright villains, especially in light of Stephen Hawkings's realistic portrayal of history as a history of stupidity. Though critical of the contemporary human condition, it seems many writers have nonetheless opted for "progress," crediting themselves with a proto-superhuman perspective. It is very tempting to love the notion of the superhero if we believe we may become superhuman ourselves.

Sadly, all this virtuous "progress" made some superheroes rather dull. *Superman* was prissy and sexless. *Captain America* was unable to become intoxicated by alcohol, so Jack Kirby became the presiding genius of a new anti-hero format for superheroes in the 1960s. The timing is no surprise. For, beginning in the 1960s and running late into the 1970s, science fiction developed a New Wave movement of writing and thinking, characterized by a high degree of experimentation in form and in content, a more literary and artistic sensibility, along with a focus on so-called soft rather than traditional hard science. This soft science included psychiatry and psychology. As quoted in Francis Booth's *Amongst Those Left: The British Experimental Novel, 1940–1980* British science fiction writer JG Ballard said, "I've often wondered why s-f shows so little of the experimental enthusiasm which has characterized painting, music, and the cinema during the last four or five decades, particularly as these have become wholeheartedly speculative, more and more concerned with the creation of new states of mind, constructing fresh symbols and languages where the old cease to be valid. . . .The biggest developments of the immediate future will take place, not on the Moon or Mars, but on Earth, and it is inner space, not outer, that needs to be explored. The only truly alien planet is Earth. In the past, the scientific bias of s-f has been toward the physical sciences—rocketry, electronics, cybernetics—and the emphasis should switch to the biological sciences. Accuracy, that last refuge of the unimaginative, doesn't matter a hoot. . . . It is that inner spacesuit which is still needed, and it is up to science fiction to build it!"

Inner space, then, and not outer space. Time for a more complex superhero. Now, superheroes had sex. They had neuroses. They behaved badly. Sometimes, they even chose to become supervillains instead, and so developed the more sophisticated superhero of the graphic novel. In later landmark publications such as Frank Miller's *Batman: The Dark Knight Returns* (1986) and Alan Moore's *Watchmen* (1986–1987), a new creative force was born. These novels confront the question of what human society might be like if science or pure chance gifted us superhero status. How complex, corrupting, and weary it all may prove.

INNER SPACE DYSTOPIA

And now there is the probability that augmented transhumans might exist within our lifespans.

It started some time ago. Prosthetic limbs, artificial heart valves, and gadgets such as pacemakers already exist to aid abilities and improve or extend an individual's life quality. British scientist Professor Kevin Warwick and his wife went partially transhuman in 2002, when they had chips and biosensors implanted into their arms. The biosensors networked to their nervous systems, which enabled them to sense each other's feelings. Professor Warwick could reportedly feel the same sensations as his wife from a different location. (Let's hope he wasn't a pain in her ass, but at least the experience may have resulted in his corrected behavior.) Many people think that projects such as Warwick's are nothing more than sideshow gimmicks, and yet Warwick has openly spoken of plans to evolve a community of fellow "cyborgs," networked, "via their chip implants to super-intelligent machines, creating, in effect, superhumans." The hope is that future tech will hugely enhance the latent potential in humans. As Warwick says, "Being linked to another person's nervous system opens up a whole world of possibilities."

Let's hope Wilson Fisk or the Joker don't hear word of Warwick's plans. The prospect of achieving superior intelligence or bodily attributes may be tempting or seem liberating, but such cybernetic tinkering could also clearly be used as a means of control. Whatever corporation (Wilson Fisk Cybernetics) controls the tech that enhances humans would be in a powerful position, wielding a huge degree of control over their human subjects. One can easily imagine the kind of scenario whose end point is some kind of situation akin to The Wacowskis' *Matrix* trilogy. Transhumans could have their implants hacked, their sensations monitored and manipulated, and be at risk of receiving unsolicited or unpleasant impulses.

In the hands of a supervillain, we might evolve from wise man to slave, *homo sapiens* to *homo servus*. And that evolution may happen sooner than you think. According to Ray Kurzweil, a US author and champion of transhumanism, in his 2005 book, *The Singularity Is Near*, by 2050 humans will become upgraded, kitted out with wetware in genetics, nanotech, and robotics. We will become essentially, says Kurzweil, a new species, with superior bodies and brains, near-immortal lifespans, and astonishing abilities, ". . . the culmination of the merger of our biological thinking and existence with our technology, resulting in a world that is still human but that transcends our biological roots. There will be no distinction, post-Singularity, between human and machine or between physical and virtual reality."

Then there's the cost of it all. Supervillains such as Fisk and the Joker always seem to know the quicker route to the loot. And the wealthy are likely to become yet more powerful and emotionally limited from those they rule when the evolution to transhumanism happens. They're more likely to afford the tech upgrade, establishing themselves as an elite "Übermensch" criminal class, forever enriching themselves as a ruling class with supernatural powers, lauding it over the "Untermensch" poor and oppressed over whom they exert economic control. Diabolical applications of enhancement tech would quickly be uncovered. Crime syndicates across the globe would give their recruits an edge over the enemy. After all, recruits with fewer bodily or mental limits would better their opponents on crime's battlefield and become dominant in a transhumanist future.

Is it wise to make such human enhancements readily available to the highest bidder? As Bob Dylan once said, "money doesn't talk, it swears." Giving ourselves superhuman abilities before we've evolved the ethical roadmaps and moral compass to wield such power is just asking for trouble from future supervillains. Human upgrades, as Stephen Hawking said, could either end in a new era of freedom, where even the most oppressed are liberated from their drudgery, or condemn the human race to permanent slavery. Technology is a double-edged sword. Beware who's wielding it.